6 -

THE BIT AND THE PENDULUM

From Quantum Computing to M Theory—The New Physics of Information

Tom Siegfried

John Wiley & Sons, Inc.

New York • Chichester • Weinheim • Brisbane • Singapore • Toronto

Published by John Wiley & Sons, Inc.
Published simultaneously in Canada

This publication is designed to provide accurate and authoritative information in regard to the subject matter covered. It is sold with the understanding that the publisher is not engaged in rendering professional services. If professional advice or other expert assistance is required, the services of a competent professional person should be sought.

Library of Congress Cataloging-in-Publication Data:

Siegfried, Tom.
     The bit and the pendulum : from quantum computing to m theory—the new physics of information / Tom Siegfried.
          p.   cm.
     Includes index.
     ISBN 0-471-32174-5 (alk. paper)
     1. Computer science.   2. Physics.   3. Information technology.
     I. Title.
QA76.S5159   1999
004—dc21   99-22275

Printed in the United States of America

10 9 8 7 6 5 4 3 2 1

Contents

Preface

In the course of my job, I talk to some of the smartest people in the universe about how the universe works. These days more and more of those people think the universe works like a computer. At the foundations of both biological and physical science, specialists today are construing their research in terms of information and information processing.

As science editor of the *Dallas Morning News,* I travel to various scientific meetings and research institutions to explore the frontiers of discovery. At those frontiers, I have found, information is everywhere. Inspired by the computer as both tool and metaphor, today's scientists are exploring a new path toward understanding life, physics, and existence. The path leads throughout all of nature, from the interior of cells to inside black holes. Always the signs are the same: the world is made of information.

A few years ago, I was invited to give a talk to a regional meeting of MENSA, the high-IQ society. I decided to explore this theme, comparing it to similar themes that had guided the scientific enterprise in the past. For it seemed to me that the role of the computer in twentieth-century science was much like that of the steam engine in the nineteenth century and the clock in medieval times. All three machines were essential social tools, defining their eras; all three inspired metaphorical conceptions of the universe that proved fruitful in explaining many things about the natural world.

Out of that talk grew this book. It's my effort to put many pieces of current science together in a picture that will make some sense, and impart some appreciation, to anyone who is interested.

Specialists in the fields I discuss will note that my approach is to cut thin slices through thick bodies of research. No doubt any single chapter in this book could easily have been expanded into a book of its own. As they stand, the chapters that follow are meant not to be comprehensive surveys of any research area, but merely to provide a flavor of what scientists at the frontiers are up to, in areas where information has become an important aspect of science.

Occasional passages in this book first appeared in somewhat different form in articles and columns I've written over the years for the *Dallas Morning News*. But most of the information story would never fit in a newspaper. I've tried to bring to life here some of the subtleties and nuances of real-time science that never make it into the news, without bogging down in technicalities.

To the extent I've succeeded in communicating the ideas that follow, I owe gratitude to numerous people. Many of the thoughts in this book have been shaped over the years through conversations with my longtime friend Larry Bouchard of the University of Virginia. I've also benefited greatly from the encouragement, advice, and insightful questions over dinner from many friends and colleagues, including Marcia Barinaga, Deborah Blum, K. C. Cole, Sharon Dunwoody, Susan Gaidos, Janet Raloff, JoAnn Rodgers, Carol Rogers, Nancy Ross-Flanigan, Diana Steele, and Jane Stevens.

I must also express deep appreciation for my science journalist colleagues at the *Dallas Morning News:* Laura Beil, Sue Goetinck, Karen Patterson, and Alexandra Witze, as well as former *News* colleagues Matt Crenson, Ruth Flanagan, Katy Human, and Rosie Mestel.

Thanks also go to Emily Loose, my editor at Wiley; my agent, Skip Barker; and of course my wife, Chris (my harshest and therefore most valuable critic).

There are in addition countless scientists who have been immensely helpful to me over the years, too many to attempt to list here. Most of them show up in the pages that follow.

But I sadly must mention that the most helpful scientist of all, Rolf Landauer of IBM, did not live to see this book. He died in April 1999, shortly after the manuscript was completed. Landauer was an extraordinary thinker and extraordinary person, and without his influence and inspiration I doubt that this book would have been written.

Tom Siegfried
May 1999

Introduction

I think of my lifetime in physics as divided into three periods. In the first period . . . I was in the grip of the idea that Everything is Particles. . . . I call my second period Everything is Fields. . . . Now I am in the grip of a new vision, that Everything is Information.

—JOHN ARCHIBALD WHEELER,
Geons, Black Holes, and Quantum Foam

John Wheeler likes to flip coins.

That's not what he's famous for, of course. Wheeler is better known as the man who named black holes, the cosmic bottomless pits that swallow everything they encounter. He also helped explain nuclear fission and is a leading expert on both quantum physics and Einstein's theory of relativity. Among physicists he is esteemed as one of the greatest teachers of the century, his students including Nobel laureate Richard Feynman and dozens of other prominent contributors to modern science.

One of Wheeler's teaching techniques is coin tossing. I remember the class, more than two decades ago now, in which he told all the students to flip a penny 50 times and record how many times it came up heads. He taught about statistics that way, demonstrating how, on average, heads came up half the time, even though any one run of 50 flips was likely to turn up more heads than tails, or fewer.*

---

*Wheeler taught a class for nonscience majors (I was a journalism graduate student at the time) at the University of Texas at Austin. In his lecture of January 24, 1978, he remarked that a good rule of thumb for estimating statistical fluctuations is to take the square root of the number of events in question. In tossing 50 coins, the expected number of heads would be 25; the square root of 25

Several years later, Wheeler was flipping coins again, this time to help an artist draw a picture of a black hole. Never mind that black holes are invisible, entrapping light along with anything else in their vicinity. Wheeler wanted a special kind of picture. He wanted it to illustrate a new idea about the nature of information.

As it turns out, flipping a coin offers just about the simplest possible picture of what information is all about. A coin can turn up either heads or tails. Two possibilities, equally likely. When you catch the coin and remove the covering hand, you find out which of the two possibilities it is. In the language that computers use to keep track of information, you have acquired a single bit.

A bit doesn't have to involve coins. A bit can be represented by a lightbulb—on or off. By an arrow, pointing up or down. By a ball, spinning clockwise or counterclockwise. Any choice from two equally likely possibilities is a bit. Computers don't care where a bit comes from—they translate them all into one of two numbers, 0 or 1.

Wheeler's picture of a black hole is covered with boxes, each containing either a zero or a one. The artist filled in the boxes with the numerals as a student tossed a coin and called out one for heads or zero for tails. The resulting picture, Wheeler says, illustrates the idea that black holes swallow not only matter and energy, but information as well.

The information doesn't have to be in the form of coins. It can be patterns of ink on paper or even magnetic particles on a floppy disk. Matter organized or structured in any way contains information about how its parts are put together. All that information is scrambled in a black hole's interior, though—incarcerated forever, with no possibility of parole. As the cosmologist Rocky Kolb describes the situation, black holes are like the Roach Motel. Information checks in, but it doesn't check out. If you drop a coin into a black hole, you'll never know whether it lands heads or tails.

But Wheeler observes that the black hole keeps a record of the information it engulfs. The more information swallowed, the bigger

---

is 5, so in tossing 50 coins several times you would expect the number of heads to vary between 20 and 30. The 23 of us in the class then flipped our pennies. The low number of heads was 21, the high was 30. Average for the 23 runs was 25.4 heads.

the black hole is—and thus the more space on the black hole's surface to accommodate boxes depicting bits. To Wheeler, this realization is curious and profound. A black hole can consume anything that exists and still be described in terms of how much information it has digested. In other words, the black hole converts all sorts of real things into information. Somehow, Wheeler concludes, information has some connection to existence, a view he advertises with the slogan "It from Bit."

It's not easy to grasp Wheeler's idea of connecting information to existence. He seems to be saying that information and reality have some sort of mutual relationship. On the one hand, information is real, not merely an abstract idea. On the other hand reality—or existence—can somehow be described, or quantified, in terms of information. Understanding this connection further requires a journey beyond the black hole (or perhaps deep inside it) to glimpse the strange world of quantum physics.

In fact, Wheeler's black hole picture grew from his desire to understand not only information, but also the mysteries of the subatomic world that quantum physics describes. It's a description encoded in the elaborate mathematical rules known as quantum mechanics.

Quantum mechanics is like the U.S. Constitution. Just as the laws of the land must not run afoul of constitutional provisions, the laws of nature must conform to the framework established by quantum mechanics' equations. And just as the U.S. Constitution installed a radically new form of government into the world, quantum requirements depart radically from the standard rules of classical physics. Atoms and their parts do not obey the mechanics devised by Newton; rather, the quantum microworld lives by a counterintuitive code, allowing phenomena stranger than anything Alice encountered in Wonderland.

Take electrons, for example—the tiny, negatively charged particles that swarm around the outer regions of all atoms. In the world of large objects that we all know, and think we understand, particles have well-defined positions. But in the subatomic world, particles behave strangely. Electrons seem to be in many different places at once. Or perhaps it would be more accurate to say that an electron isn't anyplace at once. It's kind of smeared out in a twilight zone of possibilities. Only a measurement of some sort, an observation, creates a specific, real location for an electron out of its many possible locations.

Particles like that can do strange things. Throw a baseball at a wall, and it bounces off. If you shoot an electron at a wall, it might bounce off, but it also might just show up on the other side of the wall. It seems like magic, but if electrons couldn't do that, transistors wouldn't work. The entire consumer electronics industry depends on such quantum weirdness.

Wall-hopping (the technical term is *tunneling*) is just one of many quantum curiosities. Another of the well-known quantum paradoxes is the fact that electrons (and other particles as well) behave sometimes as particles, sometimes as waves. (And light, generally thought of as traveling in waves, sometimes seems to be a stream of particles instead.) But light or electrons are emphatically not both particles and waves at the same time. Nor are they some mysterious hybrid combining wave and particle features. They simply act like waves some of the time and like particles some of the time, depending on the sort of experiment that is set up to look at them.

It gets even more bizarre. Quantum mechanics shows no respect for common notions of time and space. For example, a measurement on an electron in Dallas could in theory affect the outcome of an experiment in Denver. And an experimenter can determine whether an electron is a wave or particle when it enters a maze of mirrors by changing the arrangement of the mirrors—even if the change is made after the electron has already passed through the maze entrance. In other words, the choice of an observer at one location can affect reality at great distances, or even (in a loose sense) in the past. And so the argument goes that observers, by acquiring information, are somehow involved in bringing reality into existence.

These and other weird features of the quantum world have been confirmed and reconfirmed by experiment after experiment, showing the universe to be a much stranger place than scientists of the past could possibly have imagined. But because quantum mechanics works so well, describing experimental outcomes so successfully, most physicists don't care about how weird it is. Physicists simply use quantum mechanics without worrying (too much) about it. "Most physicists," says Nobel laureate Steven Weinberg, "spend their life without thinking about these things."[1]

Those who do have pondered the role of measurement in all these quantum mysteries. The electron's location, its velocity, and whether it's a particle or wave are not intrinsic to the electron itself, but aspects of reality that emerge only in the process of measurement. Or, in other words, only in the process of acquiring information.

Some scientists have speculated that living (possibly human) experimenters must therefore be involved in generating reality. But Wheeler and most others say there is nothing special about life or consciousness in making an "observation" of quantum phenomena. Photographic film could locate the position of an electron's impact. Computers can be programmed to make all sorts of observations on their own (kind of the way a thermostat measures the temperature of a room without a human watching it).

Nevertheless, there still seems to be something about quantum mechanics (something "spooky," as Weinberg says) that defies current human understanding. Quantum measurements do not merely "acquire" information; in some sense, they create information out of quantum confusion. To Wheeler, concrete reality emerges from a quantum fog in the answers to yes-or-no observational questions.

"No element in the description of physics shows itself as closer to primordial than the elementary quantum phenomenon, that is, the elementary device-intermediated act of posing a yes-no physical question and eliciting an answer," says Wheeler. "Otherwise stated, every physical quantity, every it, derives its ultimate significance from bits."[2] That is, It from Bit.

In his autobiography, Wheeler attempts to express this idea more simply: "Thinking about quantum mechanics in this way," he wrote, "I have been led to think of analogies between the way a computer works and the way the universe works. The computer is built on yes-no logic. So, perhaps, is the universe. . . . The universe and all that it contains ('it') may arise from the myriad yes-no choices of measurement (the 'bits')."[3]

I couldn't say whether Wheeler is on the right track with this. I asked one prominent physicist, a leading authority on traditional physics, what he thought of Wheeler's "It from Bit." "I don't know what the hell he's talking about," was the reply. But when I asked a leading authority on information physics, Rolf Landauer of IBM, I got a more thoughtful answer.

"I sympathize in a general way with this notion that handling information is linked to the laws of physics," Landauer told me. "I'm not sure I understand all the things he's saying or would agree with him. But I think it's an important direction to pursue."[4]

In a larger sense, though, whether Wheeler is right is not the big issue here. To me, it is more significant that he formulated his approach to understanding the deepest mysteries of the universe in terms of information. That in itself is a sign of the way scientists are thinking these days. Wheeler's appeal to information is symptomatic of a new approach to understanding the universe and the objects within it, including living things. This new approach may have the power to resolve many mysteries about quantum physics, life, and the universe. It's a new view of science focused on the idea that information is the ultimate "substance" from which all things are made.

This approach has emerged from a great many smart people who, like Wheeler, are all in some way engaged in trying to figure out how the universe works. These people do not all talk to each other, though. They are specialists who have fenced off the universe into various fields of research. Some study the molecules of life, some study the brain, some study electrons and quarks. Some, the cosmologists, ostensibly deal with the whole universe, but only on a scale so gross that they have to neglect most of the phenomena within it.

Most of these specialists are only partly aware of the findings at the frontiers of other fields. From the bird's-eye view of the journalist, though, I can see that many specialists have begun to use something of a common language. It is not a shared technical vocabulary, but rather a way of speaking, using a shared metaphor for conceptualizing the problems in their fields. It is a metaphor inspired by a tool that most of these specialists use—the computer.

Since its invention half a century ago, the electronic computer has gradually established itself as the dominant machine of modern society. Computers are found in nearly every business and, before long, will dwell in nearly every home. Other machines are perhaps still more ubiquitous—telephones and cars, for example—but the computer is rapidly proliferating, and no machine touches more diverse aspects of life. After all, in one form or another, computers are found within all the other important machines, from cars and telephones to televisions and microwave ovens.

Nowhere has the computer had a greater impact than in science. Old sciences have been revitalized by the computer's power to do calculations beyond the capability of paper and pencil. And, as other writers have noted, computers have given birth to a new realm of scientific study, dubbed the science (or sciences) of complexity. Complexity sciences have provided much of the new common language applied in other fields. All the hoopla about complexity is thus perhaps warranted, even if often exaggerated. Yet in any case the technical contributions of complexity science are just part of the story, the part provided by the computer as a tool. The broader and deeper development in the gestating science of the twenty-first century is the impact of the computer as metaphor. The defining feature of computing is the processing of information, and in research fields as diverse as astrophysics and molecular biology, scientists like Wheeler have begun using the metaphor of information processing to understand how the world works in a new way.

As science is pursued from the computer perspective, it is becoming clear to many that information is more than a metaphor. Many scientists now conceive of information as something real, as real as space, time, energy and matter. As Wheeler puts it, "Everything is Information." It from Bit.

This is not the first time an important technology has inspired a new view of the universe. In ancient times, Heraclitus of Ephesus taught (around 500 B.C.) that the fundamental substance in nature was fire, and that the "world order" was determined by fire's "glimmering and extinguishing." "All things are an exchange for fire, and fire for all things, as are goods for gold, and gold for goods," Heraclitus wrote.[5]

In the Middle Ages, society's most important machine was the mechanical clock, which inspired a view of the universe adopted by Isaac Newton in his vision of a mechanistic world governed by force. In the nineteenth century, the importance of the steam engine inspired a new science, thermodynamics, describing nature in terms of energy.

Four decades ago, the German physicist Werner Heisenberg

compared the worldview based on energy to the teachings of Heraclitus. "Modern physics is in some ways extremely close to the doctrines of Heraclitus," Heisenberg wrote in *Physics and Philosophy.* "If we replace the word 'fire' by the word 'energy' we can almost repeat his statements word for word from our modern point of view. Energy is indeed the material of which all the elementary particles, all atoms and therefore all things in general are made, and at the same time energy is also that which is moved. . . . Energy can be transformed into movement, heat, light and tension. Energy can be regarded as the cause of all changes in the world."[6]

Heisenberg's views still reflect mainstream scientific thought. But nowadays a competing view is in its ascendancy. Like the clock and steam engine before it, the computer has given science a powerful metaphor for understanding nature. By exploring and applying that metaphor, scientists are discovering that it expresses something substantial about reality—namely, that information is something real. In fact, I think that with just a little exaggeration, this view can be neatly expressed simply by paraphrasing Heisenberg's paraphrase of Heraclitus: "Information is indeed the material of which all the elementary particles, all atoms and therefore all things in general are made, and at the same time information is also that which is moved. . . . Information can be transformed into movement, heat, light and tension. Information can be regarded as the cause of all changes in the world."

This statement, at the moment, is more extreme than most scientists would be comfortable with. But I think it expresses the essential message, as long as it's clear that the new information point of view does not replace the old metaphors. Science based on information does not invalidate all the knowledge based on energy, just as energy did not do away with force. When the energy point of view, inspired by the steam engine, captured control of the scientific viewpoint, it did not exterminate Newtonian clockwork science. The new view fit in with the old, but it provided a new way of looking at things that made some of the hard questions easier to answer. In a similar way, the information-processing viewpoint inspired by the computer operates side by side with the old energy approach to understanding physics and life. It all works together. The information viewpoint just provides a different way of understanding and offers new insights into old things, as well as suggesting avenues of investigation that lead to new discoveries.

It is as if scientists were blind men feeling different sides of the proverbial elephant. After gathering profound understanding of nature from the perspective of the clockwork and steam engine metaphors, science is now looking at a third side of the universe. I believe that the major scientific discoveries of the next century will result from exploring the universe from this new angle.

Many scientists may still regard talk about the "reality" of information to be silly. Yet the information approach already animates diverse fields of scientific research. It is being put to profitable use in investigating physics, life, and existence itself, revealing unforeseen secrets of space and time.

Exploring the physics of information has already led to a deeper understanding of how computers use energy and could someday produce practical benefits—say a laptop with decent battery life. And information physics may shed light on the mysteries of the quantum world that have perplexed physicists like Wheeler for decades. In turn, more practical benefits may ensue. The quantum aspects of information may soon be the method of choice for sending secret codes, for example. And the most powerful computers of the future may depend on methods of manipulating quantum information.

Biology has benefited from the information viewpoint no less than physics. Information's reality has reshaped the way biologists study and understand cells, the brain, and the mind. Cells are not merely vats of chemicals that turn food into energy, but sophisticated computers, translating messages from the outside world into the proper biological responses. True, the brain runs on currents of electrical energy through circuits of cellular wires. But the messages in those currents can be appreciated only by understanding the information they represent. The conscious brain's mastery at transforming "input" from the senses into complex behavioral "output" demonstrates computational skills beyond the current capability of Microsoft and IBM combined.

Information has even invaded the realm of cosmology, where the ultimate questions involve the origin of space, time, and matter—in short, existence itself. As Wheeler's black hole drawing illustrates, information, in the most basic of contexts, is something physical, an essential part of the foundation of all reality.

There are many hints from the frontiers of research that the information viewpoint will allow scientists to see truths about existence that were obscured from other angles. Such new truths may someday offer the explanation for existence that visionary scientists like Wheeler have long sought. Wheeler, for one, has faith that the quest to understand existence will not be futile: "Surely someday, we can believe, we will grasp the central idea of it all as so simple, so beautiful, so compelling that we will all say to each other, 'Oh, how could it have been otherwise! How could we all have been so blind so long!' "[7] It could just be that the compelling clue that Wheeler seeks is as simple as the realization that information is real. It from Bit.

Tom— Again congratulations! And see you, & Lofe, at DFN! Regards — Jaw
29 III '90

INFORMATION, PHYSICS, QUANTUM

THE SEARCH FOR LINKS

John Archibald Wheeler

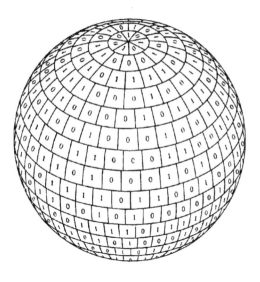

Physics Departments, Princeton University and University of Texas
February 1990 Preprint

The cover of a preprint that John Wheeler sent to me in 1990, showing his drawing of a black hole covered by "bits."

Chapter 1

Beam Up the Goulash

It's always fun to learn something new about quantum mechanics.

—Benjamin Schumacher

Had it appeared two months later, the IBM advertisement in the February 1996 *Scientific American* would have been taken for an April Fools' joke.

The double-page ad, right inside the front cover, featured Margit and her friend Seiji, who lived in Osaka. (Margit's address was not disclosed.) For years, the ad says, Margit shared recipes with Seiji. And then one day she e-mailed him to say, "Stand by. I'll teleport you some goulash."

"Margit is a little premature," the ad acknowledged. "But we're working on it. An IBM scientist and his colleagues have discovered a way to make an object disintegrate in one place and reappear intact in another."

Maybe the twenty-third century was arriving two hundred years early. Apparently IBM had found the secret for beaming people and paraphernalia from place to place, like the transporters of the famous TV starship *Enterprise*. This was a breakthrough, the ad proclaimed, that "could affect everything from the future of computers to our knowledge of the cosmos."

Some people couldn't wait until April Fools' Day to start making jokes. Robert Park, the American Physical Society's government affairs officer, writes an acerbic (but funny) weekly notice of what's new in physics and public policy that is widely distributed over the Internet. He noted and ridiculed the goulash ad, which ran not only in *Scientific American* but in several other publications, even *Rolling Stone*. He pointed out that IBM itself didn't believe in teleporting goulash, citing an article in the *IBM Research Magazine* that said "it is well to make clear at the start" that teleportation "has nothing to do with beaming people or material particles from one place to another."

"So what's going on?" Park asked. "There are several theories. One reader noted that many research scientists, disintegrated at IBM labs, have been observed to reappear intact at universities." [1]

Moderately embarrassed by such criticism, IBM promptly prepared an Internet announcement directing people to a World Wide Web page offering a primer on the teleportation research alluded to in the ad. "Science fiction fans will be disappointed to learn that no one expects to be able to teleport people or other macroscopic objects in the foreseeable future," the Web explanation stated, "even though it would not violate any fundamental law to do so." So the truth was out. Neither Margit nor IBM nor anybody else has the faintest idea how to teleport goulash or any other high-calorie dish from oven to table, let alone from orbit to Earth. That's still science fiction. But the truth is stranger still. Serious scientists have in fact begun to figure out how, in principle, teleportation might work.

The story of teleportation begins in March 1993. In that month the American Physical Society held one of its two annual meetings (imaginatively known as "the March meeting") in Seattle. Several thousand physicists showed up, most of them immersed in the study of silicon, the stuff of computer chips, or other substances in the solid state. There are usually a few out-of-the-mainstream sessions at such meetings, though, and this time the schedule listed one about the physics of computation.

Among the speakers at that session was Charles Bennett of IBM, an expert in the quantum aspects of computer physics. I had visited him a few years earlier at his lab, at the Thomas J. Watson Research

Center, hidden away in the tree-covered hills a little ways north of New York City. And I'd heard him present talks on several occasions, most recently in San Diego, the November preceding the March meeting. When I saw him in Seattle, I asked if there was anything new to report. "Yes!" he enthusiastically exclaimed. "Quantum teleportation!"

This was a rare situation for a science journalist—covering a conference where a scientist was to present something entirely new. Most "new" results disseminated at such meetings are additional bits of data in well-known fields, or answers to old questions, or new twists on current controversies. Quantum teleportation was different. Nobody had ever heard of it before. It was almost science fiction coming to life, evoking images of Captain Kirk dematerializing and then reappearing on some alien world in *Star Trek*.

In retrospect, quantum teleportation should have been a bigger story. But it isn't easy to get new developments in quantum physics on the front page of the newspaper. Besides, it was just a theoretical idea in an obscure subfield of quantum research that might never amount to anything, and it offered no real hope of teleporting people or even goulash. To try to make teleportation a news story would have meant playing up the science-fiction-comes-to-real-life aspect, and that would have been misleading, unwarranted sensationalism, or so I convinced myself. Instead I wrote about quantum teleportation for my weekly science column, which runs every Monday in the science pages tucked away at the back of Section D. My account appeared on March 29, the same day the published version of the research appeared in the journal *Physical Review Letters*. So if teleporting goulash ever does become feasible, March 29, 1993, will be remembered as the day that the real possibility of teleportation was revealed to the world. (Unless, of course, you'd prefer to celebrate on March 24, the day that Bennett presented his talk in Seattle on how to teleport photons.)

Teleporting Information

Almost two years later (and a year before the IBM goulash ad appeared), Samuel Braunstein, a quantum physicist at the Weizmann Institute in Israel, was asked to present a talk to the science-fiction club in Rehovot. What better topic, he thought, than quantum tele-portation? News of this idea hadn't exactly dominated the world's media in the time since Bennett had introduced it in Seattle. But teleportation had attracted some attention among physicists, and the science-fiction connection provided a good angle for discussing it with the public.

Braunstein immediately realized, though, that talking about tele-portation presented one small problem—it wasn't exactly clear what "teleportation" really is. It's no good just to say that teleportation is what happens when Scotty beams Kirk back up to the *Enterprise*. So Braunstein decided he had to start his talk by devising a teleportation definition. "I've seen *Star Trek*," he reasoned, "so I figure I can take a stab at defining it."[2]

In the TV show, characters stood on a "transporter" platform and dissolved into a blur. They then reformed at their destination, usually on the surface of some mysterious planet. To Braunstein, this sug-gested that teleportation is "some kind of instantaneous 'disembod-ied' transport." But hold the phone. Einstein's laws are still on the books, and one of them prohibits instantaneous anything (at least whenever sending information is involved). Therefore, Braunstein decided, teleportation is just "some kind of disembodied transport." That's still a little vague, he realized, and it might include a lot of things that a science-fiction club surely didn't have in mind. A fax, for example, transports the images on a sheet of paper to a distant lo-cation. And telephones could be thought of as teleporting sound waves. In both cases, there is a sort of disembodied transport. But neither example is really in harmony with the science-fiction sense of teleportation.

Teleporting, Braunstein decided, is not making a copy of some-thing and sending the copy to somewhere else. In teleportation, the original is moved from one place to another. Or at least the original disintegrates in one place and a perfect replica appears somewhere else. A telephone line, on the other hand, merely carries a copy of sound waves, emitted and audible at point A, to a receiver at point B, where the sounds are regenerated. A fax machine spits the origi-nal sheet out into a waiting basket as a copy appears at some distant location. The original is not teleported—it remains behind.

But perhaps copying of some sort is involved in "real" teleportation, Braunstein suggested. Maybe *Star Trek*'s transporters work like a photocopy machine with too strong a flashlamp, vaporizing the original while copying it. The information about all the object's parts and how they are put together is stored in the process and then sent to the planet below. The secret of teleportation, then, would lie not in transporting people, or material objects, but in information about the structure of whatever was to be teleported.

Somehow, then, the *Star Trek* teleporter must generate blueprints of people to be used in reconstructing them at their destination. Presumably the raw materials would be available, or perhaps the original atoms are sent along and then reassembled. In any case, the crew members vaporized on the transporter platform magically rematerialize into the same people because all the information about how those people were put together was recorded and transported.

Naturally this process raises a lot of questions that the script writers for *Star Trek* never answered. For example, just how much information would it take to describe how every piece of a human body is put together?

They might have asked the U.S. National Institutes of Health, which plans to construct a full 3-D model of the human body (computer-imaged to allow full visualization of all body parts, of course), showing details at any point down to features a millimeter apart. Such a model requires a lot of information—in terms of a typical desktop computer, about five hard drives full (at 2 gigabytes per hard drive). Maybe you could squeeze it all into a dozen CD-ROMs. In any case, it's not an inconceivable amount for a computer of the twenty-third century, or even the twenty-first.

But wait. The NIH visible human is a not a working model. In a real human body, millimeter accuracy isn't good enough. A molecule a mere millimeter out of place can mean big trouble in your brain and most other parts of your body. A good teleportation machine must put every atom back in precisely its proper place. That much information, Braunstein calculated, would require a billion trillion desktop computer hard drives, or a bundle of CD-ROM disks that would take up more space than the moon. And it would take about 100 million centuries to transmit the data for one human body from one spot to another. "It would be easier," Braunstein noted, "to walk."

So the information-copying concept did not seem very promising for teleportation, although the hang-up sounds more like an engineering problem than any barrier imposed by the laws of physics. Technically challenging, sure. But possible in principle.

Except for one thing. At the atomic scale, it is never possible to obtain what scientists would traditionally consider to be complete information. Aside from the practical problems, there is an inherent limit on the ability to record information about matter and energy. That limit is the Heisenberg uncertainty principle, which prohibits precise measurement of a particle's motion and location at the same time. Heisenberg's principle is not a mere inconvenience that might be evaded with sufficient cleverness. It expresses an inviolate truism about the nature of reality. The uncertainty principle is the cornerstone of quantum mechanics.

Quantum mechanics codifies the mathematical rules of the subatomic world. And they are not rules that were made to be broken. All the consequences predicted by quantum mathematics, no matter how bizarre, have been confirmed by every experimental test. Quantum mechanics is like Perry Mason—it never loses. And there is no court of appeal. So if quantum mechanics says you cannot physically acquire the information needed to teleport an object, you might as well give up. Or so it would seem. But in the decade of the 1990s, physicists have learned otherwise. You may not be able to teleport ordinary information. But there is another kind of information in the universe, concealed within the weirdness of quantum mechanics. This "quantum information" can be teleported. In fact, it is the marriage of information physics to quantum weirdness that makes teleportation possible, even if it's not quite the sort of teleportation that *Star Trek*'s creator, Gene Roddenberry, had in mind.

So when the IBM ad writers mentioned objects that could already be teleported, they referred not to goulash or even anything edible, but to the most fundamental pieces of reality: objects described by the mathematics of quantum mechanics.

Quantum Objects

Understanding quantum objects is like enjoying a Hollywood movie—it requires the willing suspension of disbelief. These objects are nothing like rocks or billiard balls. They are fuzzy entities that elude concrete description, defying commonsense notions of space and time, cause and effect. They aren't the sorts of things you can hold in your hand or play catch with. But they are important objects nonetheless—they could someday be used to decipher secret military codes, eavesdrop on sensitive financial transactions, and spy on confidential e-mail. And as the IBM ad suggested, the study of quantum objects could transform the future of computers and human understanding of the universe.

Typical quantum objects are the particles that make up atoms—the familiar protons and neutrons clumped in an atom's central nucleus and the lightweight electrons that whiz around outside it. The most popular quantum objects for experimental purposes are particles of light, known as photons. A quantum object need not be a fundamental entity like a photon or electron, though. Under the right circumstances, a group of fundamental particles—such as an entire atom or molecule—can behave as a single quantum object.

Quantum objects embody all the deep mysteries of quantum mechanics, the most mysterious branch of modern science. Part of the mystery no doubt stems from the name itself, evoking the image of an auto repairman who specializes in a certain model of Volkswagen. But in quantum physics the term *mechanics* refers not to people who repair engines, but to the laws governing the motion of matter, the way classical Newtonian mechanics describes collisions between billiard balls or the orbits of the planets.

It is not easy to understand quantum mechanics. In fact, it's impossible. Richard Feynman put it this way: "Nobody understands quantum mechanics."[3] Niels Bohr, who understood it better than anybody (at least for the first half of the twentieth century) expressed the same thought in a slightly different way, something to the effect that if quantum mechanics doesn't make you dizzy, you don't get it. To put it in my favorite way, anybody who claims to understand quantum mechanics, doesn't.

To the extent that scientists do understand quantum mechanics, explaining it would require a book full of a lot of very sophisticated math. Many such books have already been written. Unfortunately, they don't all agree on the best math to use or how to interpret it. It might seem, then, that understanding quantum mechanics and quantum objects is hopeless. But in fact, if you don't worry about the details, quantum mechanics can be made ridiculously simple. You just have to remember three basic points: Quantum mechanics is like money. Quantum mechanics is like water. Quantum mechanics is like television.

Quantum Money

Historically, the first clue to the quantum nature of the universe was the realization that energy is quantized—in other words, energy comes in bundles. You can't have just any old amount of energy, you have to have a multiple of the smallest unit. It's like pennies. In any financial transaction in the United States the amounts involved have to be multiples of pennies. In any energy transaction, the amounts involved must be measured in fundamental packets called quanta.

Max Planck, the German physicist who coined the term *quantum* (from the Latin for "how much"), was the first to figure out this aspect of energy. An expert in thermodynamics, Planck was trying to explain the patterns of energy emitted by a glowing-hot cavity, something like an oven. The wavelengths of light emitted in the glow could be explained, Planck deduced, only by assuming that energy was emitted or absorbed in packets. He worked out the math and showed that the expectations based on his quantum assumption were accurately fulfilled by the light observed in careful experiments.

By some accounts, Planck privately suggested that what he had found was either nonsense or one of the greatest discoveries in physics since Newton. But Planck was no revolutionary. He tried to persuade himself that energy packets could merge in flight. That way light could still be transmitted as a wave; it had to break into packets only at the point of emission by some object (or absorption by another). But in the hands of Albert Einstein and Niels Bohr, Planck's quanta took on a life beyond anything their creator had intended. Einstein proposed that light was composed of quantum particles in flight, and he showed how that idea could explain certain features of the photoelectric effect, in which light causes a material to emit electrons. Bohr used quantum principles to explain the architecture of the atom. Eventually it became apparent that if energy were not like money, atoms as we know them could not even exist.

Quantum Water

Planck announced the existence of quanta at the end of 1900; Einstein proposed that light was made up of quantum particles in 1905; Bohr explained the hydrogen atom in 1913. Then followed a decade of escalating confusion. By the early 1920s it was clear that there was something even weirder about quantum physics than its monetary aspect—namely, it was like water.

How in the world, physicists wondered, could Einstein be right about light being made of particles, when experiments had proven it to be made of waves? When they argued this point over drinks, the answer was staring them in the face (and even kissing them on the lips). Ice cubes. They are cold, hard particles, made of water. Yet on the oceans, water is waves.

The path to understanding the watery wave nature of quantum physics started in 1925 when Werner Heisenberg, soon to become the father of the uncertainty principle, had a bad case of hay fever and went off to the grassless island Heligoland to recover. Isolated from the usual distractions, he tried out various mathematical ways of describing the motion of multiple electrons in atoms. Finally one evening he hit on a scheme that looked promising. He stayed up all night checking his math and finally decided that he'd found a system that avoided all the previous problems. As morning arrived, he was still too excited to sleep. "I climbed up onto a cliff and watched the sunrise and was happy," he later reported.[4]

Unwittingly, Heisenberg had reinvented a system of multiplication using arrangements of numbers called matrices. Only later, when he showed the math to his professor at the University of Göttingen, Max Born, was he told what kind of math he had "invented." "Now the learned Göttingen mathematicians talk so much about . . . matrices," Heisenberg told Niels Bohr, "but I do not even know what a matrix is."[5]

Heisenberg's version of quantum mechanics was naturally designated matrix mechanics. It treated quantum objects (in this case, electrons) as particles. The next year, the Austrian physicist Erwin Schrödinger, in the midst of a torrid love affair, found time to invent another description of electrons in atoms, using math that treated each electron as a wave. (Schrödinger's system was creatively called wave mechanics.) As it turned out, both wave and matrix methods produced the same answers—they were mathematically equivalent—even though they painted completely different pictures of the electron.

Immediately physicists became suspicious. For years, Einstein had argued that light was made of particles, despite all the evidence that light was wavy. Now along comes Schrödinger, insisting that electrons (thought since 1897 to be particles) were really waves. And even before Schrödinger had produced his paper, the first experimental evidence of wavy electrons had been reported.

Bohr, who had merged quantum and atomic physics years earlier, was the first to devise an explanation for the double lives led by electrons and photons. In a famous lecture delivered in 1927, he presented a new view of reality based on a principle he called complementarity. Electrons—and light—could sometimes behave as waves, sometimes as particles—depending on what kind of experiment you set up to look at them.

Suppose you have an ice bucket containing H_2O. You don't know whether the water is in the form of liquid or ice cubes, and you're not allowed to look into the bucket. So you try an experiment to find out. You grab a pair of tongs and stick them into the bucket, and out come ice cubes. But if you dip in a ladle, you get liquid water.

An obvious objection arises—that the bucket contains cubes floating in liquid. But you try another experiment—turning the bucket upside down. Some of the time, ice cubes fall out, with no liquid. But on other occasions only liquid water spills out. Water, like electrons, can't seem to make up its mind about how to behave. Quantum physics seems to describe different possible realities, which is why quantum mechanics is like television.

Quantum Television

TV signals ride on invisible waves. Emanating from transmitter towers or bouncing off of satellites in orbit, those waves carry comedy, drama, sports, and news from around the world into your living room every moment of the day. But you can't hear or see or smell these waves. TV's sights and sounds are just possibilities, not fully real until you turn the TV on and bring one of those intangible prospects to life.

Physicists view the subatomic world of quantum physics in a similar way. Atoms and smaller particles flutter about as waves, vibrating not in ordinary space but in a twilight zone of different possibilities. An electron is not confined to a specific orbit about an atom's nucleus, but buzzes around in a blur. The math describing that electron's motion says not where it is, but all the places it might be. In other words, the electron is simultaneously in many places at once, or at least that is one way of interpreting what the math of quantum mechanics describes. You can think of it as an electron in a state of many possible realities. Only when it is observed does the electron assume one of its many possible locations, much the way punching the remote control makes one of the many possible TV shows appear on the screen.

Another way of interpreting the math is to say that the quantum description of nature is probabilistic—that is, a complete quantum description of a system tells nothing for certain, but only gives the odds that the system will be in one condition or another. If you turn the quantum ice bucket upside down, you can predict the odds that it will come out liquid or cubes—say, 7 times out of 10 cubes, 3 times out of 10 liquid. But you can't predict for sure what the outcome will be for any one try.

In a similar probabilistic way, for something as simple as the position of an electron, quantum mechanics indicates the odds of finding the electron in any given place. Quantum physics has therefore radically changed the scientific view of reality. In the old days of classical physics, Newton's clockwork universe, particles followed predictable paths, and the future was determined by the past. But in the description of physical reality based on quantum mechanics, "objects" are fuzzy waves. The object we know as something solid and tangible might show up here, might show up there. Quantum math predicts not one future, only the odds for various different futures.

This is clearly the way life is for atoms and electrons. You simply cannot say, or even calculate, where an electron is in an atom or where it will be. You can only give the odds of its being here or there. And it is *not* here or there until you look for it and make a measurement of its position. Only then do you get an answer, and all the other possibilities just disappear. This is a difficult point to grasp, but it needs to be clear. It's not just a lack of knowing the electron's "real" position. The electron does not *have* a real position until somebody (or some*thing*) measures it. That's why there is no avoiding the Heisenberg uncertainty principle. You can't measure an electron's velocity or position, just a range of possibilities.

It's this probabilistic aspect of quantum mechanics that makes quantum information so mysterious, and so rich. And it's an important part of the reason why quantum information makes teleportation possible.

Quantum Information

Everyday life is full of examples of information carriers—the electromagnetic waves beaming radio and TV signals, electrons streaming along telephone and cable TV lines, even quaint forms of information like ink arranged to form words and pictures. But whatever its form, information can be measured by bits and therefore described using the 1s and 0s of computer language. Whether information comes in the form of ink on paper, magnetic patterns on floppy disks, or a picture worth 10,000 kilobytes, it can always be expressed as a string of 1s and 0s. Each digit is a bit, or binary digit, representing a choice from two alternatives—like a series of answers to yes-no questions, or a list of the heads-or-tails outcomes of repeatedly tossing a coin.

Quantum information, on the other hand, is like a spinning coin that hasn't landed. A quantum bit is not heads or tails, but a simultaneous mix of heads and tails. It is a much richer form of information than, say, Morse code, but it is also much more fragile. Quantum information cannot be observed without messing it up—just like you can't see whether a coin is heads or tails while it is still in the air. Looking at quantum information destroys it. In fact, it is impossible to copy (or "clone") a quantum object, since making a copy would require measuring the information, and measurement destroys.

So you could not "fax" a copy of a quantum object—sending the information it contains to a distant location and retaining a copy for yourself. But in teleportation, as Braunstein pointed out, the original does not remain behind. There is no copy, only a disintegration and then reconstruction of the original. Perhaps a quantum object could be moved from one location to another without the need to measure (and thereby destroy) the information it contains. And that is exactly the strategy that Charles Bennett described at the March 1993 physics meeting in Seattle.

Quantum Teleportation

At the meeting, Bennett presented the paper for an international team of collaborators.* They had produced an intricate scheme for teleporting a quantum object. More precisely, they figured out how to convey all the quantum information contained by a quantum object to a distant destination, without transporting the object itself. It was kind of like the quantum equivalent of sending the *Encyclopaedia Britannica* from New York to Los Angeles, leaving the bound volumes of paper in New York and having only the ink materialize on fresh paper in L.A. Accomplishing this trick requires the sophisticated application of quantum technicalities. But to keep it as simple as possible, think of a quantum object as carrying information by virtue of the way that it spins. Say in this case the object in question is a photon, a particle of light. When unobserved, the photon's spin is a mix of multiple possibilities, like all those possible TV channel signals streaming through your living room. You could picture the photon as spinning around an axis pointing in many different directions. (It would be as if the Earth's North Pole could point toward many different constellations at once. For the sake of navigators everywhere, it's a good thing that such multiple quantum possibilities are not detectable in systems the size of a planet.)

---

*They included Gilles Brassard and Claude Crepeau of the University of Montreal, Richard Jozsa of the University of Plymouth in England, Asher Peres of the Israel Institute of Technology in Haifa, and William Wootters of Williams College in Massachusetts.

Anyway, measuring the photon freezes its spin in one of those directions, or states, destroying all the other possibilities. The destruction of all those different possible spin directions means a lot of information gets lost. In other words, a pure photon contains information about many directions; a measured photon contains information about only one direction. The trick of teleportation is to communicate all the "possibility" information contained in a "pure" photon that has not yet been observed. In other words, you can use teleportation to send a message without knowing what the message is.

So suppose Bennett's favorite scientist, Bob at IBM, wants to study a photon produced in Los Angeles. He wants Alice, his colleague at UCLA, to send him all the information about that pure photon's condition. This request poses a serious challenge for Alice. If she so much as looks at that particle she will obliterate most of the information it contains. She needs to find a way of sending information about the particle's pristine condition to Bob without herself knowing what that condition is. Clearly, this is a job for quantum teleportation.

Bennett and his colleagues reasoned that the quantum information of the pure photon could be teleported if Alice and Bob had prepared properly in advance. The preparation scheme requires an atom that will emit twin photons and a way to send one of those photons to Bob and the other to Alice. They then save these twins for later use. These photon twins have a peculiar property—one of them "knows" instantly if something happens to the other. In other words, a measurement of one of them instantaneously affects the other one. If Alice were to sneak a peek at her photon and find that its spin pointed up, Bob's photon would immediately acquire a spin pointing down.

This ethereal twin-photon communication is at the heart of many quantum-mechanical mysteries, and it was the feature that Einstein singled out as evidence that quantum mechanics was absurd. Einstein never imagined that it would instead be the evidence that quantum mechanics could be practically useful. These twin photons are today known as EPR photons in honor of Einstein and his two collaborators, Boris Podolsky and Nathan Rosen, who introduced the twin photons in 1935 in a paper challenging quantum mechanics.

Einstein and colleagues believed that they could show quantum mechanics to be an incomplete theory of nature. When the quantum equations were applied to two particles that had interacted (or were produced from a common source, as when an atom spits out two photons), a paradox seemed to arise. The exact equation describing photon B depends on whether anyone has measured photon A. So before photon A's spin is measured, photon B's spin could turn out to be either up or down. But once the spin of photon A is measured, photon B's spin is instantly determined, no matter how far away it is from photon A. So a single measurement instantly determines the spin of the two photons; neither has a precisely determined spin before the measurement.

Einstein found this situation unreasonable, but Bohr replied that there was no inconsistency in the quantum picture. If you analyzed how the experiment had to be conducted, you would see that in no actual case would a paradox arise. True, photon B could turn out to have either spin up or spin down, but no experimental arrangement could get both answers—only one or the other.

Einstein conceded that the quantum description was consistent, but he still didn't like it. Nevertheless, by the 1980s real-life experiments showed that Bohr was right. The EPR twin particles do in some weird way share information about each other.

It's that shared information that makes EPR twins useful for sending quantum information. In advance, Alice and Bob prepare a pair of EPR photons; Alice keeps one and Bob takes the other. Then at any later time Alice can simply put another photon—the one Bob wanted to know about—in a box with her EPR photon. Then she measures the whole thing. She then sends that result to Bob via fax or e-mail. Bob can use that information to transform his EPR twin into an exact replica of the photon that Alice put into the EPR box. Voilà. Teleportation.

Okay, so the original photon went nowhere. It's still in Alice's box with her EPR twin. But notice that Alice can no longer determine the pure photon's original condition. Once the particles mixed in the box, the original photon gave up its information to the mixture. The original photon cannot be reconstructed; in effect, it has been destroyed. But Bob now possesses a photon with all the information contained in Alice's original. The information has been teleported, and that's all that matters. If this is how *Star Trek* does it, Captain Kirk materializes on an alien planet out of a bunch of new atoms—the original atoms are left behind on the transporter floor on the *Enterprise*. But all the information that makes Kirk Kirk is contained in his new body.

Of course, it's important to keep in mind that Kirk is fictional, and so is teleporting people. But quantum teleportation is real. In the fall of 1997, a paper appeared on the Internet reporting a successful laboratory teleportation of a photon along the lines that Bennett and his colleagues had described. In December the paper was published in the well-known British scientific journal *Nature,* and quantum teleportation then made headlines, even showing up on the front pages of some papers, playing off the *Star Trek* analogy.

Scientists are still a long way from sending goulash to Osaka. But it may not be too long before teleporting quantum information has some practical applications. Most likely, those applications will be in computers of the future. It might be desirable, for instance, to transfer quantum information from one part of a computer to another, or even from one computer to another. Even sooner than that, though, quantum information might have an important practical application of use to the government, the military, spies, and banks—sending secret messages. Quantum teleportation may still be mostly science fiction, but quantum cryptography is real.

Quantum Cryptography

Actually, quantum cryptography predates quantum teleportation. The first in-depth discussions of using quantum information to send secret codes began to appear in the 1980s. By the time I visited Bennett in 1990, he had a working quantum cryptography operation in his office, with computers exchanging messages between one end of a table and the other. This would hardly do for a Moscow-Washington hotline. But it wasn't long before quantum messages were zooming through optical fiber cables over substantial distances. By the mid-1990s quantum cryptography was a serious subject of study.

The study of cryptography itself, of course, has a long history, going back well before anybody ever heard of quantum physics.

Ancient Greeks and Romans used various systems for secret codes. (Julius Caesar used one that shifted letters by three places in the alphabet, so that the letter *B*, for instance, would be represented by *E*.) But before the twentieth century, cryptography was more a concern of generals and ambassadors than of scientists. In recent decades, though, secret codes have been given serious scientific scrutiny—thanks largely to the electronic computer and the math of information theory. Twentieth-century studies have established that the problem of sending a secure message boils down to giving the two communicating parties a decoding key available only to them, as Bennett explained to me during my visit. "The problem of cryptography can be reduced to the problem of sharing a secret key," he said.[6] If two communicators share a key, one of them can use it to encode a message; the other uses the same key to work backward and decode the message.

A key can be as simple as a string of random digits, and nothing is more convenient than simply using 0s and 1s. (That will make it easy on the computers, which always prefer to speak in the binary language of 0s and 1s.) So suppose our favorite communicators, Alice and Bob, both possess copies of a key containing a long string of 0s and 1s, in random order. Alice can then encode a message to Bob simply by altering the random numbers of the key according to particular rules. Bob can decode Alice's message by comparing the bits Alice sends him to the bits in the key.

For example, the coding rule could be that if the digit of the message matches the digit of the key, write down a one. If it doesn't match, write down a zero. Like this:

Alice sends:	0 1 0 0 1 0 0 1
Key says:	0 1 1 0 0 1 1 1
Bob writes:	1 1 0 1 0 0 0 1

By agreeing to alter the random numbers of the key in a particular way, Alice and Bob can code a message that can be interpreted only with the secret key. The message itself could be sent over an open channel; an eavesdropper would not be able to understand the message without the key.

Of course, the key can be used only once. A clever eavesdropper could detect patterns in the messages if the same key is used over and over, possibly helping to break the code. So to be perfectly safe, a key needs to be discarded after use. (You could still send several messages, though, if your key contained a long list of digits and you needed to use only a small portion of them per message.)

The remaining problem is how to share the secret key. A floppy disk filled with random bits works pretty well, if Alice and Bob can first meet face-to-face for a disk exchange. But if they live far apart and have lousy travel budgets, how can they exchange a key and be sure that it remains secret? One of them could mail a key to the other, or even use FedEx. But there would always be a chance that a clever spy could intercept the key and copy it. Alice and Bob's communication would not be certain to be secure.

But quantum information is spyproof. If Alice transmits a key to Bob using quantum information, they can be absolutely sure that no eavesdropper has copied it. "What quantum cryptography can do is to allow the two parties to agree on that random secret without ever meeting or without ever exchanging a material object," Bennett explained.

Alice and Bob could compile a list of 0s and 1s by sending photons that are oriented in one direction or another by passing them through filters like Polaroid sunglasses. (A polarized filter blocks out light that is not oriented in the desired way.) In one quantum cryptography approach, Alice can send Bob photons oriented in various ways. She might send one horizontally, the next vertically, the next diagonally—tilted either to the left or the right. Bob, however, has to make a choice of how to try to detect Alice's photon. He can choose to distinguish vertical from horizontal, or left-tilting from right-tilting.

With this setup Alice can send bits to Bob using this code:

Vertical photon = 1
Horizontal photon = 0
Right-tilting = 1
Left-tilting = 0

Now, suppose Alice sends Bob a vertical photon through their quantum communications channel (optical fiber, maybe). Bob can then tell Alice through an open channel (e-mail or telephone) whether he was looking for vertical-horizontal photons or tilted photons. If he was looking for tilted photons, he won't get the right answer, so Alice and Bob agree to throw out that photon. But if he was looking for horizontal or vertical photons, he'll know it was vertical. They both write down a 1. Note that the beauty of this scheme is that Bob doesn't have to tell Alice what bit he received, only that he was set up properly to receive what Alice sent. Only they know that it was a vertical photon, representing a 1.

By repeating this process, after a while they can build up a long string of 0s and 1s to use as a code key.

But what about the possibility of an eavesdropper (call her Eve) who taps into the optical fiber and intercepts Alice's photons? Eve could not do so undetected. Even if she sent the photon on to Bob after intercepting it, her observation would have disrupted the information it contains. Bob and Alice merely need to check a string of their bits every once in a while to make sure they are free from errors. If Eve has been listening in, about a fourth of the bits that Bob receives will not be what they should be.* If Alice and Bob *do* detect the presence of an eavesdropper, they can invoke a system for throwing away some of the bits they send. Eve could then get only some of the information in their messages, since she does not know which bits are kept and which are discarded.

But even if Eve knows only some of the bits, she might be able to break the code. So Alice and Bob have to go through another process, similar to techniques used in correcting errors in computer codes, to create a series of random numbers that Eve cannot possibly duplicate. Alice and Bob could then use that series of random numbers as a key and then communicate merrily away in code using ordinary e-mail.

Today quantum cryptography is a lively research field, taken seri-

---

*If Eve is lucky enough to set her filter in the right way, she will intercept the bit successfully. But some of the time her filter won't match Alice's, as she has no way of knowing which filter Alice will use. In some of those cases Bob will receive an intercepted photon from Eve that doesn't match what Alice originally sent, even though Bob and Alice have their filters set in the same way. By checking for such errors, Bob and Alice could discover Eve's nefarious plot.

ously enough to attract research funding from the U.S. military. In fact, quantum cryptography is considered by some experts to be the best long-term solution to recent assaults on the nation's current cryptography systems.

The best current secret-code systems are based on hard-to-solve mathematical problems, as fans of the Robert Redford film *Sneakers* already know. These problems really are hard—even harder than calculating the salary cap for an NBA basketball team's payroll. They involve numbers more than 100 digits long, with the problem being to conquer the number by dividing it. The divisors must be whole numbers that leave no remainder, and they must have no such divisors of their own. Purdue University mathematician Samuel Wagstaff compares the problem to the TV game show *Jeopardy*. You get the answer and have to figure out the question. If the answer is 35, the question is "What is 5 times 7?" Note that 5 and 7 are what mathematicians call prime numbers, because they have no whole number divisors (other than 1 and themselves). Finding the primes that multiply to make a big number is the key to breaking today's toughest secret codes.

Obviously, 35 would not be a good number to use for a code, since the primes that produce it have just been revealed in the last paragraph. So code makers instead build their codes on enormously long numbers. You can't decipher such a coded message unless you know the prime numbers (called factors) that produce the long number. As recently as the early 1990s, a number with 120 digits or so was considered uncrackable. But lately computer collaborations have succeeded in dividing and conquering much longer numbers. In 1997 an international team headed by Dr. Wagstaff broke down a 167-digit number into its two factors—one with 87 digits, the other with 80.

Code makers did not immediately panic, because the 167-digit number had special properties that made it easier to solve than other numbers that length. But code numbers may not be safe for long. Improvements in computer speed and factoring methods may soon push the safety limit above 200 digits.

Therefore interest in quantum cryptography has grown, and so has technical progress toward practical systems. In fact, at the Los Alamos Laboratory in New Mexico, researchers reported in 1997 that they had transmitted quantum-coded messages between two computers (named Alice and Bob) through 48 kilometers (about 30 miles) of underground optical fiber. Researchers at the University of Geneva in Switzerland have reported similar success. Los Alamos physicist Richard Hughes says that the ability to send photon codes through the air, without the aid of optical fibers, has also been demonstrated over short distances. That raises the prospect of sending quantum-safe secret messages to and from communication satellites.

All this has brought quantum information a long way from its obscure standing at the beginning of the 1990s. When I had talked to Bennett at his office in 1990, he was unsure that quantum cryptography would ever go commercial and seemed pessimistic about the future usefulness of other aspects of quantum weirdness for practical purposes. But by the 1993 meeting in Seattle, it seemed more likely that quantum mechanics permitted doing things that nobody thought possible outside the realm of cinematic special effects. So I asked Bennett there whether quantum physics held other surprises in store.

"I would guess there are probably a few more surprises," he said. "The quantum mechanics that we've had around for all these years is a very strange and beautiful theory that appears to describe our world in a very accurate way. But also it has consequences which I think we have not entirely discovered yet."[7]

He was right. And many of those consequences stem from an appreciation of the peculiar properties of quantum information.

Quantum Information Theory

If you've been paying close attention, you'll have noticed a subtle discrepancy in the stories of quantum teleportation and quantum cryptography. Teleportation sends a special kind of information—quantum information—from one location to another. Quantum cryptography's ultimate goal is sending ordinary ("classical") information, or bits.

Classical information is well described mathematically by what scientists call "information theory," invented half a century ago and widely used in applications from telephone transmissions to computer memory storage. But Alice and Bob's photons carry quantum information. Their quantum cryptography scheme produces classical information in the end, but to get there you have to transmit quantum information back and forth. The math of classical information theory won't be sufficient, therefore, to describe quantum cryptography (or teleportation). Describing these quantum magic tricks requires something that didn't exist a decade ago—quantum information theory.

Like Sandburg's fog on little cat feet, quantum information theory crept into existence in the 1990s, almost unnoticed outside the small group of specialists who were trying to figure out how to apply quantum physics to the study of information. Among the pioneers of this obscure discipline is Benjamin Schumacher, a physicist at Kenyon College, a small liberal arts school in Gambier, Ohio. At a meeting in Dallas in October 1992, Schumacher gave a short talk presenting the concept of a bit of quantum information, calling it a qubit (pronounced CUE-bit).* It was more than just a clever name. The concept of a qubit made it possible to study quantum information quantitatively, much the way that classical information theory made it possible to measure quantities of ordinary information.

One of the key uses of information theory is in helping computer designers to make efficient use of information. In the early days of computers, storage memory was expensive. A computer program needed to do its job using the least possible amount of information. Messages to be stored using scarce memory resources had to be condensed as much as possible to take up the minimum possible space. The math of information theory made it possible to calculate how a given message could be stored using the least possible number of bits.

_____

*Later, Schumacher told me how he and William Wootters (one of the teleportation paper authors) had discussed quantum information during a visit by Wootters to Kenyon in May 1992. On the way back to the airport, Schumacher said, they laughed about it. "We joked that maybe what was needed was a quantum measure of information, and we would measure things in qubits was the joke, and we laughed. That was very funny. But the more I thought about it, the more I thought it was a good idea. . . . I thought about it over the summer and worked some things out. It turned out to be a really good idea."

Schumacher's qubits could be used in a similar way to compute the most efficient possible compression of quantum information.

As far as I can tell, nobody jumped onto the qubit bandwagon immediately. But by June of 1994, qubits had become more than a bit player in the computing physics field. In that month I drove to Santa Fe for a workshop on quantum physics and information, the sort of event where that small band of quantum information theorists would gather to bring each other up to date. It was at that meeting, I think, that everyone involved realized that quantum information theory had become something important.

"Quantum information theory's a pretty new game," Schumacher told me during lunch at Sol y Sombra, a ranch on Santa Fe's outskirts where the workshop was held. "Five years ago there wasn't such a thing. Now there is, sort of."[8]

Bennett was also at the meeting, and he concurred that quantum information was coming of age. "Nobody was working on this a few years ago," he said. "Now we're really able to make progress in understanding . . . all the different ways in which classical and quantum information can interact, in what cases one can be substituted for another, and how many bits or qubits you need to transmit certain kinds of messages."[9]

A few months before the Santa Fe workshop, *Nature* had published a short account of Schumacher's theorem on qubits. His actual paper had not appeared in print yet, but it had been circulated among scientists on the Internet and was recognized by many as an important step toward understanding quantum information more fully. An article in *Nature* was a sign to people outside the clique that quantum information might be worth taking seriously.

At the workshop, I sat through various talks extolling the usefulness of quantum information and quizzed many of the participants about it during lunch and breaks. Several of the scientists expressed excitement over the prospect that quantum information theory could someday solve deep quantum mysteries. Seth Lloyd, then a physicist at the Santa Fe Institute and now at MIT, expressed the hope that the weirdness of quantum phenomena might seem more reasonable when viewed with the insight provided by qubits.

"Quantum mechanics is weird," Lloyd said. "That's an experimental fact. Everybody finds it weird. . . . It involves features that are hard to understand." He recalled Niels Bohr's belief that anyone not made dizzy by quantum mechanics didn't understand it. Quantum information theory might help scientists to "explore the strange aspects of quantum mechanics without becoming dizzy," Lloyd said. "These different ideas about . . . how information can be processed and used in a purely quantum mechanical fashion are good ways of figuring out just where the weirdness in quantum mechanics lies."[10] Schumacher also believed that profound insights would emerge from quantum information theory. He cited in particular the peculiar property that quantum information could not be copied. (Copying requires measuring, and measuring, you'll remember, destroys quantum information.) The impossibility of copying a quantum object had been proved in the early 1980s by William Wootters and Wojciech Zurek, in what they called the "no-cloning theorem." It is easier to clone a sheep than a quantum object. Schumacher thinks the impossibility of copying quantum information might point to explanations for many quantum mysteries, including the inviolability of the Heisenberg uncertainty principle.

But by far the greatest source of excitement at the Santa Fe workshop stemmed from another recent development, a paper from Bell Labs suggesting that quantum information might someday be truly useful in the operation of a new kind of computer.

It was not a computer that anyone yet knew how to build—it existed only in theory. And computers that exist only in theory have a name of their own, after the theoretical computer conceived before real computers existed: the Turing machine.

Chapter 2

Machines and Metaphors

The "machine" has been a powerful metaphor for several centuries, but the idea of the machine has changed over time. The mechanical, clocklike, image has been superseded by new technical inventions shaping the images, such as the steam engine, electricity and, recently, electronics. . . . Needless to say, these shifts in the real world of technology have also changed our images of the world through our machine metaphors. At the heart of this modern metaphor stands the computer.

—ANDERS KARLQVIST AND UNO SVEDIN,
The Machine as Metaphor and Tool

Alan Turing was a terrible typist.

Or so suggests Turing's biographer, Andrew Hodges, who allows that at least some of the sloppy typing was the fault of a cat named Timothy who liked to paw at the keyboard. In any case, Turing had a certain fascination with typewriters. And in the end, the allure typewriters had for Turing contributed to their demise.

Turing, one of the most brilliant mathematicians of his era, became famous in some circles for cracking the German secret code during World War II. Later he analyzed the physics of pattern formation, and that work is still invoked today to explain mysteries like how tigers got their stripes. Nowadays his name is perhaps best known in connection with the "Turing test," a sort of game designed to decide whether a computer is truly intelligent.[1]

But Turing's most important and lasting legacy was figuring out how digital computers could work in the first place. In the 1930s, even before the modern digital computer had been invented, Turing figured out the basic principles underlying any computing machine. He hadn't set out to invent the computer—just to analyze mechanical methods of solving mathematical problems. Without any real computers around already, he had to imagine a mechanical way of computing in terms of some familiar machine. He chose the typewriter.

Turing was far from the first scientist to draw inspiration from a machine. Machines have often shown scientists the way to discovering deep principles about how the world works.

"Many of the most general and powerful discoveries of science have arisen, not through the study of phenomena as they occur in nature, but, rather, through the study of phenomena in man-made devices, in products of technology," the engineer John R. Pierce has noted. "This is because the phenomena in man's machines are simplified and ordered in comparison with those occurring naturally, and it is these simplified phenomena that man understands most easily."

Typewriters are rather simple machines, and Turing understood them. He transformed that understanding into the intellectual basis for the machine that replaced them. Nowadays, in its turn, the computer inspires scientists in a similar way as they apply the metaphor of computing to the scientific study of both life and the physical universe. In short, the computer has led science to a new conception of reality. To understand how this happened, it helps to see how machines have changed ideas about reality in the past.

As Pierce emphasized, the computer is not the first instance of a machine-based view of the natural world. The grandest example of such machine-science symbiosis was the Newtonian view of the universe inspired by the mechanical clock. A more recent—and in fact, still current—example involves the laws of energy, born in the study of the steam engine.

Superparadigms

The clock, steam engine, and computer have all inspired metaphorical frameworks for science that I like to call superparadigms. To me (and as far as I know, I made this word up), a superparadigm is a point of view about what's ultimately fundamental in determining what happens in the world. I mean something a little grander and more general than the specific paradigms made famous by the science historian Thomas Kuhn. He viewed paradigms as frameworks of thought (or "disciplinary matrices") within which researchers in a given field practiced "normal science." A superparadigm underlies and unites the various restricted paradigms guiding research in specific disciplines.

For example, Newton's superparadigm described the universe in terms of motion governed by force, the way the moving parts of a clockwork mechanism were driven by the pull of weights attached to ropes. That basic mechanistic view provided a convenient picture for understanding why things happen and how things change—whether apples falling on people's heads or planets and moons eclipsing one another. That basic idea of force thus formed a foundation on which other science could be built in the Newtonian spirit.

Throughout the eighteenth century, all the paradigms of normal science were built on this Newtonian foundation. In the nineteenth century, a superparadigm based on the steam engine superseded Newton's notion of force. In the twentieth century, the computer superseded—or perhaps engulfed—both preceding superparadigms.

Now it would be foolish to contend that this view reflects a reasoned historical analysis of the depth provided by Kuhn or any other science historian. It's merely a cartoon picture of the history of science that emerges from my efforts to understand what's going on in science today. But there are certain aspects of what went on in centuries past that do illuminate the current fascination with computers and information as the way of understanding life and the universe. Things happen faster now than they did in the last century, or the Middle Ages, but they nevertheless seem to be some of the same things. Basically, it appears, the most important machine of its time captures the imagination of the entire culture, and science cannot escape that cultural influence in the way it pursues its attempts to explain the world.

That was surely the case with the mechanical clock. Today, the notion of a "clockwork" universe is synonymous with Newton's physics, and people sometimes speak as though Newton's science created the clockwork metaphor for the universe. But that idea was around long before Newton spelled out his laws of motion. Newton's science did not create the clockwork metaphor, but exploited it.

The Clock

Clocks had already been an important part of culture for a long time. And the idea of connecting clocks to astronomy goes back many centuries. During the darkest of Europe's dark ages, Chinese artisans built elaborate water clocks that represented astronomical phenomena. And the Roman orator Cicero once compared regular heavenly motions to the regularities of timepieces. But the specific use of the clockwork metaphor for the universe didn't really catch on until somebody invented the weight-driven mechanical clock, sometime in the late thirteenth century.

Nobody knows who did it, but most historians guess that some monk wanted a better way to tell when it was time to pray. After all, even the Rolex of sundials was no help at night. Water clocks could take a licking and keep on dripping, but not at a steady enough rate to make sure prayers were always punctual. So some genius figured out how to connect toothed wheels and dangling weights into a contraption with mechanical movements regular enough to be used as clockwork.

Though its inventor evidently neglected to secure the patent rights, the mechanical clock quickly captured the time-telling market. Clocks were hot commodities by 1320 or so, when another medieval genius, named Nicole Oresme, was born somewhere in northern France. During his lifetime, communal clocks began to appear in most towns of any significant size. Just as today's teenagers have grown up with computers, Oresme, educated at the University of Paris before turning to life as a churchman (eventually becoming a bishop), came of age in an epoch dominated by the clock.[2] It's not surprising that when he contemplated the universe, clocks were in his thoughts.

Oresme was one of the first to transform such thoughts into writing. The earliest specific discussion of the clock-universe metaphor I can find is in one of his many books, *Le Livre du ciel et du monde* (The book of the sky and the world), a translation of, and commentary on, some of the works of Aristotle. "The heavenly bodies move with such regularity, orderliness, and symmetry that it is truly a marvel; and they continue always to act in this manner ceaselessly," Oresme observed. "Summer and winter, night and day they never rest."[3] A few pages later, he adds: "When God created the heavens, He put into them motive qualities and powers just as He put weight and resistance against these motive powers in earthly things. . . . The powers against the resistances are moderated in such a way, so tempered, and so harmonized that the movements are made without violence; thus, violence excepted, the situation is much like that of a man making a clock and letting it run and continue its own motion by itself."[4]

Even in Oresme's day, three centuries before Newton, clock-makers were already crafting intricate timepieces designed specifically to mimic the motions of the heavenly bodies. An early Italian clock by Giovanni de Dondi, for example, incorporated a planetarium-like series of wheels depicting the movement of sun, moon, and planets. The Strasbourg Cathedral clock, finished in 1354 and then rebuilt in 1574, was a famous example of a device illustrating the mechanical movements of the heavens. The rebuilt Strasbourg clock may even have inspired Descartes, foremost among the originators of the mechanistic view of nature. Descartes was one of those giants with broad shoulders on whom Newton stood to see the way to construct the details of the clockwork universe.

In his famous *Principia* of 1687, Newton transformed the metaphor of clockwork into something scientifically more tangible. He called it force. Just as the moving parts of a clockwork mechanism were driven by the pull of weights attached to ropes, forces directed the motion of matter throughout the universe. Newton's force prevailed as the central concept of physics for a century and a half.

By the end of the eighteenth century, though, a new machine had encroached on the clock's dominance in European culture. It was a machine eventually to become almost as important to society as the clock, and therefore a machine with the metaphorical power to alter the Newtonian worldview. The steam engine.

The Steam Engine

The steam engine story actually begins in 1698, when Thomas Savery, an English merchant and inventor, patented a steam device for pumping water out of coal mines. It was not a very good machine. But a few years later Thomas Newcomen struck a business deal with Savery. Savery had the patent, but Newcomen had a better idea for how to make steam engines useful. Dozens of Newcomen engines were built in England in the decades to follow—perhaps more than a thousand by the end of the eighteenth century.

Though a commercial success, Newcomen's engine was far from the most energy-efficient way to get power from steam. The big breakthrough in steam engine technology came in 1765, when James Watt, going for a walk on Easter Sunday, suddenly realized that a different design could harness steam power more effectively. Watt was an instrument maker at the University of Glasgow, interested not in discovering laws of nature but in finding a way to improve the Newcomen engine in order to make money. Finally he hit on a design that could make the steam engine transform much more of its available energy into useful work.

It took Watt a few years to patent his idea and then many more years of design manipulation and financial struggle before his version of the steam engine began to take over the Industrial Revolution. But soon the steam engine began to proliferate the way the mechanical clock had in Oresme's century. By the opening of the nineteenth century, Watt-style steam engines had become the preeminent source of power in industrial Britain, not only pumping water out of mines to permit better production of the coal needed to fuel the revolution, but also driving flour and textile mills and serving as the source of force for the nascent British factories. Soon the steam engine also became a driving force in transportation. Just as the clock had become the dominant machine of medieval society, the steam engine became the symbol of the Industrial Revolution.

All along, engineers worked to improve Watt's engine, but there was little science available to guide them. No one really knew what fundamental principles, what laws of nature, dictated the behavior of steam engines the way that Newton's laws dictated the motion of the planets. But for a machine so important, such ignorance was not likely to be tolerated for long. The world was waiting for Carnot.

Born into a prominent French family in 1796, Sadi Carnot was a brilliant youth, and at sixteen he enrolled in the Ecole Polytechnique, where he studied geometry, mechanics, and chemistry. He became a military engineer, turning to the study of the steam engine after getting out of the army. He was apparently quite impressed with the steam engine's emergence as the prime mover of the industrial revolution in England, and he saw that France lagged far behind in the development of industrial power. The steam engine, Carnot noted, had revived the dying coal mining industry in England and provided that island with no end of uses—milling grain, forging iron, driving ships, spinning and weaving cloth, and transporting the heaviest of loads.

And yet, in spite of its importance and decades of advances in its design, the steam engine (or heat engine, in Carnot's terms) had not been understood fully in terms of basic scientific principles. "The theory of its operation is rudimentary," Carnot asserted, "and attempts to improve its performance are still made in an almost haphazard way."[5]

In 1824 Carnot published a treatise on the steam engine, designed to rectify this lack of theoretical understanding and to generate some enthusiasm among his countrymen for developing steam power. It has become one of the most famous works in the history of science, titled *Reflexions on the Motive Power of Fire*. In it Carnot spelled out the physical principles underlying the workings of the steam engine and in so doing identified general principles that constrained the operation of any heat engine. Using Carnot's principle, it was possible to calculate the maximum amount of work that any heat engine could produce. The ultimate limitations were not in the design of any device, but inherent in the way nature worked.

At first, nobody noticed. Carnot was shy, perhaps a bit misanthropic, and evidently not much of a self-promoter. And then in 1832 he died young during a cholera epidemic. The world was thereby deprived of many more years of his genius, and, due to the customs of the times, it was even deprived of much of what he had done while alive—most of his papers and effects were burned.

Fortunately, his *Reflexions* were rediscovered. A decade after its publication, the French engineer Emile Clapeyron revived Carnot's ideas, and in the late 1840s Carnot's work gained the widespread appreciation it warranted—thanks to the British physicist William Thomson, later to become Lord Kelvin. Through the work of Thomson and others, the lessons of the steam engine taught science the laws of thermodynamics—laws that still stand today, untouched by the revolutions of twentieth-century physics.

The parallel with the mechanical clock is unavoidable. In the same way that the clock was exploited by Newton, the steam engine led to another revolution in science. Clocks became the culturally supreme machine and inspired a metaphor for describing the universe, leading to the new science of Newtonian physics. Steam engines became society's prime mover and inspired the new science of thermodynamics, in the process producing a new metaphor—the universe as a heat engine. (Scientists began to wring their hands over the "heat death of the universe," the logical deduction from thermodynamics that the universe was a like a big engine that would eventually run out of steam.)

The thermodynamic description of nature supplanted Newton's central concept of force with a new supreme feature—energy. Thermodynamics—the study of energy and its transformations—was applied to all sorts of things. Physics, chemistry, the processes of life, and thus all the workings of the physical world came to be viewed in terms of energy.

As the twentieth century ends, this view of reality is changing. Energy is still common scientific currency, a viable superparadigm for explaining what happens when things change. But more and more another point of view competes with, and sometimes supersedes, the worldview of the steam era. Nowadays information is the currency most widely accepted around the scientific world. The computer— only half a century old in its electronic form—has clearly become society's dominant machine, and information has become science's favorite superparadigm. So you could sum up my cartoon history of modern science like this:

Clock

- Dominant tool in society
- Tool and metaphor for science leading to new science of Newtonian mechanics
- Metaphor for scientific worldview based on *force*

Steam Engine

- Dominant tool in society
- Object of scientific study leading to new science of thermo-dynamics
- Metaphor for scientific worldview based on *energy*

Computer

- Dominant tool in society
- Tool for science and object of scientific study leading to new science of physics of computation
- Metaphor for scientific worldview based on *information*

The Computer

The computer's history is no less intriguing than that of the clock and steam engine, if for no other reason than it's impossible to name its inventor without causing a lot of controversy. Most people have no idea who invented the electronic computer, and those who think they know can't seem to agree with one another.

In 1996, festivities marking the computer's fiftieth birthday cele-brated ENIAC (Electronic Numerical Integrator and Computer), the electronic behemoth officially put into service at the University of Pennsylvania in 1946. Its inventors were John Mauchly and J. Pres-per Eckert, with help from a large supporting cast. But those celebra-tions evoked protests from fans of a physicist named John Atanasoff. In 1973, Mauchly and Eckert had lost a patent case in which a judge declared Atanasoff to be the inventor of the electronic computer.

Atanasoff, who taught at what is now Iowa State University, had built an electronic computer prototype in 1939 (with the help of graduate student Clifford Berry) and had discussed his ideas with Mauchly. But Atanasoff soon abandoned his idea, and his machine went nowhere. Meanwhile, Mauchly joined forces with Eckert to produce the ENIAC during World War II. They continued in the computer business and were widely regarded by people in the business as the modern computer's inventors.

Eventually, though, patent issues arose that ended up in court, where Mauchly and Eckert lost the case. A decade later, Joel Shurkin, a Pulitzer Prize–winning journalist, reviewed the trial testimony and other evidence in preparing a book on the history of the computer, called *Engines of the Mind*. Shurkin concluded that the judge's decision may have been correct as a matter of law, but bunk as history. Atanasoff seems to have had better lawyers than Mauchly and Eckert, and much relevant information was never entered into the court record. "Who invented the modern computer?" wrote Shurkin. "J. Presper Eckert and John Mauchly."[6]

More specifically, ENIAC was the first large-scale general-purpose electronic digital computer. But it was not the first device to calculate. The computer owes its origin to a long history of imaginative thinkers who wanted to mechanize the computational power of the mind. To find the real inventor of mechanical calculation, we should set the Wayback machine for seventeenth-century France to visit Blaise Pascal.[7] Pascal, born in 1623, was a precocious teenager who loved math, especially geometry. Somehow he got the idea to study objects shaped like an ice cream cone—a sugar cone with a pointed bottom, not the flat-bottomed kind. Imagine turning it upside down on a tabletop. You are now looking at what mathematicians would call a cone.

Now take a sharp knife and slice the tip off the cone through a plane parallel to the table. The fresh-cut surface (or "section") traces out a circle. Now cut through the cone at an angle, and the section becomes an ellipse. Or try a vertical cut, slicing a side off the cone down to the table. This section is an open curve known as a hyperbola. The mathematics of the most important curves of nature can all be described as sections of a simple cone.

All this is explained in elementary geometry books today. But

Pascal figured it out for himself, describing these relationships and many more complicated theorems about conical geometry in a treatise written when he was merely sixteen. Descartes, the most prominent French mathematician of the day, could not believe a teen could do such sophisticated work.

But Pascal was no ordinary teen. As a young child he had proved many of Euclid's geometric theorems, even before he had ever heard of Euclid. Once his father presented the twelve-year-old Blaise with a copy of Euclid's *Elements,* a future of mathematical greatness seemed assured.

It didn't quite work out that way, though. Blaise grew from a genius child into a humorless religious fanatic as an adult. In fact, his reflections on religion (*Pensées*) made him more famous than his mathematics. He suffered throughout his life from the pain of a bad digestive system, rotten teeth, and recurring headaches, and died at thirty-nine. E. T. Bell wrote (in *Men of Mathematics*) that Pascal was "perhaps the greatest might-have-been in history."[8]

Pascal did manage some significant accomplishments. He helped Fermat, another great mathematician, to establish the foundations of probability theory. The motivation in that case was to assist a French aristocrat in his gambling strategies. And Pascal also went down in history as the world's first hacker. At age nineteen he invented a mechanical computing device that he modestly called the Pascaline.

The genesis of the Pascaline involves a certain amount of political intrigue. Ultimately, you could say that it owed its existence to the French king's councilor, Cardinal Richelieu. Pascal's father, Etienne, was no fan of Richelieu and was in danger of ruination from a political purge. But Richelieu, a fan of the stage, was entranced by a performance of Blaise's actress sister, Jacqueline. Richelieu asked Jacqueline what he could do for her, so she suggested he might decide against sending her dad to the guillotine. Richelieu complied, and in fact did even better, offering Etienne a nice political appointment as a tax collector in Normandy.

It was a tough job, though. Blaise, devoted son that he was, saw that dad was overburdened with all the calculations. So he invented the Pascaline to help out. It wasn't much as computers go—it could only add and subtract, using gears similar to those in today's car odometers. But it was a first—a novel thing in nature, as Pascal's other sister, Gilberte, wrote, "to have reduced to a machine a science which resided wholly in the mind."

Pascal's machine did not inspire an industry. It was expensive and hard to repair. (By some accounts, Blaise himself was the only licensed repairman.) It couldn't multiply. And it was worthless for spreadsheets, word processing, or playing Tetris, the sorts of things that made the personal computer popular in the twentieth century.

Some improvements in the idea came along later. Gottfried Wilhelm von Leibniz, the German philosopher-mathematician and rival of Newton, produced an improved version that could multiply. But it's not obvious that Leibniz's machine ever worked very well. The first really successful calculating machine didn't come along until 1820, when Charles Xavier Thomas de Colmar invented the arithmometer, which could add, subtract, multiply, and divide.

None of these devices approached anything like the modern idea of a computer, though. Computing is more than calculating. A general-purpose computer can read different sets of instructions to perform different sorts of tasks, and can save what it needs to remember to do those tasks. The first person with a clear vision of such a general-purpose computer was Charles Babbage, perhaps an even greater might-have-been than Pascal.

Babbage, son of a banker, was an insightful mathematician and imaginative inventor. But his political skills would make Jimmy Carter look like the Great Communicator. Babbage was a troublemaker. He didn't like the way English universities (specifically, Cambridge) clung to Newton's archaic notation for calculus, while the rest of the world used the much more sensible system developed by Leibniz. So Babbage and friends organized a campaign that eventually displaced Newton's notation. Later in life, Babbage published diatribes against the noisy annoyances of London's street musicians, earning him the response of even louder music outside his home and occasional rocks tossed through his windows.

Though Babbage charmed many of his social acquaintances, he didn't always make a great impression. The author Thomas Carlyle, for example, described Babbage as "eminently unpleasant" with his "frog mouth and viper eyes, with his hide-bound, wooden irony, and the acridest egotism looking through it."[9]

Well, you can't please everybody. Babbage, unfortunately, was never quite able to please himself, either. His grand schemes for computing machines never fully materialized, largely because he kept getting a "better idea" before any one machine was finished. And he botched negotiations with the government for continued funding. Consequently the world had to wait an extra century to see the first general-purpose computer.

Babbage was born in 1791, during the era that Watt's steam engine was rising in popularity and importance. By 1820 Babbage was a well-known mathematician with a deep interest in astronomy. In discussions with his friend the astronomer John Herschel, Babbage realized that inaccuracies in the mathematical tables astronomers used might be reduced if the numbers could be calculated by a machine. He imagined a powerful machine—powered by steam, naturally—that could compute and print out numbers for such needs as nautical charts. He designed such a machine—he called it the Difference Engine—and then sought, and got, funding from the government to begin its construction.

The Difference Engine, however, succumbed to mismanagement and cost overruns (and Babbage's inability to get along with the engineer hired to build it). But that didn't matter, because along the way Babbage came up with one of his better ideas, a machine that could do more than compute specific numbers for tables, a machine that could compute anything. Babbage tried to imagine a machine that could both calculate and store numbers, and he found for his model (the way Turing used the typewriter) the Jacquard loom, a French invention (in 1802) that had become popular in the British weaving industry. The Jacquard loom could weave any variety of patterns based on information recorded in a set of punched cards (the forerunners of the computer-punched cards of the 1950s and '60s). Babbage conceived of a similar approach to be used in a computing machine he called the Analytical Engine.

Babbage's friend Ada, Countess of Lovelace (the daughter of Lord Byron), took deep interest in the Analytical Engine and even translated an account of it that had been written in Italian. Just as punched cards regulate the patterns on fabric in a Jacquard loom, she wrote, Babbage's Analytical Engine could conduct all varieties of abstract algebra. "The Analytical Engine weaves algebraical patterns just as the Jacquard loom weaves flowers and leaves," Ada wrote.[10]

Of course, Babbage's Analytical Engine was never built. Some parts of it were made, but Babbage kept thinking up new ways of doing things that prevented progress. Furthermore, other people were coming up with good new mathematical ideas that Babbage wanted to incorporate in his machine. Especially striking was a new way of representing logical propositions mathematically, just the sort of thing that an all-purpose computer needed. That new logic came from an Englishman living in Ireland, a self-taught genius named George Boole.

Boolean Logic

Nowadays, computers are logically powerful enough to beat the world's best chess player—an astounding feat for a machine that knows just two numbers, 0 and 1. Sure, the computer can manipulate larger numbers, but ultimately all the numbers a computer computes are represented as strings of 0s and 1s.

People, on the other hand, insist on using a ten-digit system of numbers. That's not because there is something superior or more natural about the decimal system, of course; it's an accident of the biology of human hands. Computers are like organisms whose primitive ancestors had only two fingers. Their two-symbol approach to math is called the binary system.

Counting in binary is simple enough. From 0 to 1, then 10, then 11, 100, 101. But there is more than just that simplicity that makes the binary way better for performing logical processing on a computer. It has nothing to do with the nuances of electronic technology or the way vacuum tubes and transistors work. The 0 and 1 approach to computing logic was invented almost a century before electronic computers existed.

The inventor, Boole, wanted to show that the laws of human thought could be expressed by algebraic equations. Those equations, he imagined, would describe logical relationships, and solving the equations would provide the correct logical answer to a problem. Just as Babbage wanted to reduce tedious, mindless calculating to mechanical motions in a machine, Boole wanted to reduce the richness of human logical thought to algebra. Modern computers, in a way, are the offspring of a marriage between Babbage's mindless mechanism and Boole's "thoughtful" equations.

Boole made considerably more progress in realizing his ambitions than Babbage did. Boole's book *An Investigation of the Laws of Thought,* published in 1854, made a major impression. Half a century later the mathematician-philosopher Bertrand Russell acclaimed Boole as the discoverer of "pure mathematics."[11]

In his famous book, Boole worked through his ideas in rather tedious detail, but their essence isn't hard to summarize briefly. Boole reasoned that thought could be represented mathematically by assigning algebraic symbols (like x, y, z) to various concepts. Say x represents men and y represents all white things. Then the equation x minus y would refer to all men other than white men. Using this approach, Boole believed, symbols for various concepts could be manipulated mathematically to compute logical conclusions.

He worked out a set of rules for such computations. One such rule stated that the product of two symbols refers to all the individuals satisfying the definition of both symbols. So if x stands for men and y stands for white things, the product x times y would be equal to all white men. But then he had to consider the case of the product of a symbol and itself. If x is all men, then the product xx (or x squared) would also represent all men. Therefore x squared is equal to x.

Most people would have given up at this point. Making x squared equal to x sounds like a pretty illogical conclusion from a system that's supposed to define the nature of logic. But Boole realized that this rule captured an interesting insight. There are numbers for which the equation x times x equals x does hold true. In fact, there are precisely two such numbers—0 and 1. There you have it. If you want to reduce logic to equations, you better do it using only 0 and 1—in other words, you need to use the binary system.

Boole's approach sounds simple enough, but it gets complicated fast when putting a lot of symbols through a complicated chain of algebraic calculations. Some intermediate equations might even express a combination of symbols that could not be interpreted at all. For example, if x represents things producing pleasure and y stands for things producing pain, x plus y is hard to interpret. Because some things produce both pleasure and pain (so I've been told), the sum of x plus y could be less than the whole of the parts. Boole solved this conundrum by declaring that only the ultimate output of the calculations needed to be in a form that could be interpreted. Intermediate steps in the math did not have to have a clear logical meaning. "The validity of a conclusion arrived at by any symbolical process of reasoning, does not depend upon our ability to interpret the formal results which have presented themselves in the different stages of the investigation," Boole wrote. There is no problem, he said, with "the chain of demonstration conducting us through intermediate steps which are not interpretable, to a final result which is interpretable."[12]

In other words, seemingly absurd mathematical expressions could in fact lead the way to useful and valid conclusions. Today that spirit is at work in new studies trying to understand the physics of information.

Unlike Babbage, who was largely forgotten, Boole's work became well known, at least among people who studied the mathematics of logic. And in the 1930s, among those who did was Alan Turing.

Turing Machines

When scientists gather to discuss theories about computing, you don't have to wait long for them to start talking about Turing machines. It's one of those code phrases that scientists like to use to keep outsiders confused. But in this case, it's not so confusing if you remember that a Turing machine is like a typewriter.

Okay, in some ways it's a little more complicated than an ordinary typewriter, but in some ways it's simpler. In any event, Turing had a model of a typewriter in mind back in the 1930s when he came up with the whole idea.

Turing, born in 1912 in London, grew up near Hastings on the shore of the English Channel—his parents often absent because of his father's job as a civil servant in India. Alan attended King's College at Cambridge and crossed the ocean to Princeton to get his Ph.D. But before leaving Cambridge for the United States, he finished the job of destroying the fondest dreams of mathematical philosophers.

In 1931, the math world was stunned when the logician Kurt Gödel announced his proof that no system of math (at least, no system as complicated or more so than ordinary arithmetic) could be both internally consistent and complete. In other words, you could find true statements that could not be proven to be true. Mathematical scholars had dreamed of packaging all of math into a nice system of axioms and proofs that would encompass all mathematical knowledge. Gödel informed them that that they should forget about it.

A related question remained unanswered, though. Given any number of mathematical problems, could a guaranteed method be devised for deciding the answers? When the Cambridge professor M.H.A. Newman lectured on this issue, called the decidability question, Turing was inspired to investigate it. The task amounted to a search for an automatic way of determining the truth of any proposition. Put another way, finding a machine that, using some guaranteed method, could figure anything out.

A guaranteed method, Turing realized, had to be a clear, definable method, a set of specific instructions (nowadays known as an algorithm). No creative leaps of imagination or intuitive insights were allowed. The process had to be mechanical. Turing therefore had to conceive of a general-purpose computing machine before such a machine existed, and then figure out the fundamental principles behind how such a machine worked. Imagine Carnot in the same situation, designing the steam engine without ever having seen one and then figuring out the laws of thermodynamics. The Turing machine was perhaps the most ingenious imaginary invention in history.

Of course, it was part of Turing's genius to realize that solving any possible problem automatically, or "mechanically," would require a machine. He was not the first to think about a machine that reduced arithmetic to "mechanical rules." But those before him had not given much thought to what such a machine would actually look like. Turing wanted to imagine an actual mechanism. It had to be a mechanism that could deal with symbols, of course, since symbols were the mathematical way of representing things, of abstracting specific cases into general cases for all-purpose calculating. And to be reliable the machine would have to respond in the same way to the same situation. All these qualities, Turing saw, were present in the typewriter.

So in conceiving the computer, Turing had typewriters on his mind, along with some modifications and embellishments. "He imagined machines which were, in effect, super-typewriters," wrote Andrew Hodges in his biography of Turing.[13]

Turing's mutant typewriter, like any ordinary typewriter, needed to be able to make a mark on a piece of paper. He envisioned the paper as a long strip divided into squares. The paper could be pushed forward or backward, one square at a time, the way a space bar or backspace key moves a typewriter carriage.

Turing required an additional feature that only showed up later in real typewriters—an erasing key that could remove the mark from a square. Thus at its most basic, a Turing machine contained a marking device (or "read-write head") that could move forward or backward, one square at a time, and then make a mark, erase a mark, or do nothing, depending on rules for its behavior (the program).

Programming this device is simply a matter of telling the read-write head what to do after it examines one of the squares. For example, if the square in view displays a mark, the program could direct the head to move one square forward and make another mark. Or go backward two squares and change nothing. More sophisticated processing was possible by allowing the device to change "modes" the way a typewriter might be in lowercase mode or uppercase mode engaged by a shift key. Using a device like this, Turing showed, it is possible to compute anything that can be computed.

The trick is designing the supertypewriter so it could carry out a series of logical processes, step by step, the way computer programs work today. Turing showed that you could construct a machine to perform whatever steps were needed to solve a particular problem. For another kind of problem, you might need another machine, programmed differently. Any such machine is known today as a Turing machine.

In principle, you could build an infinite number of different Turing machines to compute solutions to different problems. But Turing also saw that you could write a set of instructions for imitating any of those machines. In other words, one computer could be programmed to do what any other computer can do (kind of the way IBM clones running Windows can imitate Macintoshes). Therefore the principles of computation are independent of any specific computing device (similar, in a way, to Carnot's idea that the same laws apply to all heat engines, no matter what their design). Consequently, Turing realized, it would be possible to build a "universal" computer that could simulate the workings of any other computer. For any given task, the universal computer might be slower or need more memory, but in principle it could duplicate the computations of whatever particular computer model you happened to have. Such a computer is now called a universal Turing machine.

The marvel of Turing's imaginative invention should not obscure an important issue—namely, the answer to the question he was originally investigating. Were all mathematical questions decidable? The answer, Turing's investigations showed, was no. There were some problems that even a universal Turing machine could not answer.

Turing wrote a paper outlining these ideas, "On Computable Numbers,"[14] in 1936. About the same time word arrived in England of very similar conclusions from the Princeton logician Alonzo Church. Church derived his result much differently, though, focusing on a strictly mathematical analysis. Church developed a system of mathematical logic for expressing algorithms, while Turing's approach was physical, analyzing the actual mechanical implementation of algorithms. In any case, the two approaches were equivalent, as Turing later showed.

Consequently the central insight into the nature of computing is today known as the Church-Turing thesis: A Turing machine can compute anything that can be computed.[15] The Church-Turing thesis raises serious issues for science. It seems to impose limits on what humans can compute about the world. Or, as Turing seems to have thought, it imposes limits on the world itself. Put another way, the question becomes whether computing is about mathematics or about physics. Is information processing a mathematical exercise or a physical process?

I think the answer to the last question is clear. Just as the machine-inspired metaphors of the past have produced physical concepts like force and energy, the computer metaphor has driven the development of a new form of physical science. Information is more than a metaphor. Information is real. Information is physical.

Chapter 3

Information Is Physical

Until recently, information was regarded as unphysical, a mere record of the tangible, material universe, existing beyond and essentially decoupled from the domain governed by the laws of physics. This view is no longer tenable.

—WOJCIECH ZUREK,
"Decoherence and the Transition from Quantum to Classical"

Edward Fredkin thinks that the universe is a computer.

He doesn't mean that the way the universe works reminds him of a Macintosh or that the virtual worlds of computer games are accurate simulations of reality. He insists that the universe *is* a computer simulation. All the things we see, know, and do are merely shadows in the software of some unfathomable computer. It's like the holodeck on the starship *Enterprise* in *Star Trek: The Next Generation.* All the action on the holodeck is generated by a computer in another part of the ship. The computer that drives our universe, says Fredkin, is "somewheres else."

"It's very easy to understand," he tried to persuade me when I interviewed him at a conference in Dallas. Consider a sophisticated flight simulator, he argued. A powerful computer makes it seem as though the airplane is really flying, with realistic scenery and flight-like motions controlled by a computer-driven apparatus. You can appear to fly from Dallas to Houston in such a simulator. But the computer itself goes nowhere.

"You can fly all over the country or all over the world and keep looking out the window and saying, 'Where is the computer? Where is the computer?' You won't see it because it's somewheres else. . . . In the simulated world, that computer's not there." Therefore, he concludes, a computer running a program to produce our universe can't be in our universe. "Our world is the information process that's running in that machine," he said. "But the machine isn't in our world."[1]

As far as I can tell, nobody qualified to judge him really agrees with Fredkin about this. The consensus response is essentially "Yeah, right." Although he is a legitimate physicist, and has published some interesting and worthwhile research, his views on the universe-as-computer are too far out even for *Star Trek* fans. In a column I wrote in 1993, I suggested that the chances that he is right are about the same as the chances that the Buffalo Bills will ever win the Super Bowl. So far, so good. The Bills have never won.

But they have made some of the games interesting. And as I also said back then, science doesn't have to be right to be interesting. Fredkin's ideas may be wrong, but they may also suggest insights into how the universe works anyway. He may take his ideas to extremes, but they nevertheless are based on a much less contentious principle—that the universe can in fact be described in terms of information processing.

After all, even before computers, processing information was a common feature of the universe. Brains do it all the time. All communication is just information processing. And to Fredkin, information processing is at the foundation of everything that happens in nature. Consider the momentum of a particle in flight. In traditional mechanistic physics, where the world is described as matter in motion, momentum is a fundamental conserved quantity, the product of a particle's mass and velocity. But from Fredkin's point of view, momentum is nothing more than information. It's the information that tells a particle at one point in space and time where it should be at the next moment.

All of nature's activities can be described as information processing, Fredkin insists. "It's like you have the seed of an oak tree—the information with the design of the oak tree is in there. If you process it with an environment of sunshine and rain, you eventually get an oak tree," he says. And a particle with a particular momentum gets to wherever it is going. The future comes about because a computation transforms information from the present into the new conditions representing the next instant of time.

In other words, information is the foundation of reality. It from Bit.

So even though other physicists reject Fredkin's supercomputer running "somewheres else," many embrace the idea of information as something real, something important in the description of nature. Information not as something abstract and intangible, but information as something physical.

This is not the sort of thing you learn in school. Textbooks stick to the old-guard party line describing the world in terms of matter and energy. The truth is, most scientists still view the world that way. But during the twentieth century a handful of physical science pioneers have encountered a new world that the old vocabulary is unable to describe.

These pioneers began with a tiny "demonic" clue that something was amiss with the standard matter-energy description of nature. Then came the invention of a language to describe information with mathematical precision. Next was the widespread use of the computer, inspiring research that led to deep new physical insights into the relationship between energy and information processing. All of this led to the inescapable conclusion that information is an essential concept for describing the world. "The computer has made us aware," says Rolf Landauer, "that information is a physical entity."[2]

Information's reality is essential to understanding the fundamental physics of computers. But that's not all. Information is everywhere. Information is behind the science of the telephone company's clear long-distance connections, TV transmissions, signals from space probes, the making and breaking of secret codes, the sounds on compact disks, and the special effects in major motion pictures. All of these technologies exploit the reality of information.

Today this realization that information is real is at the heart of a new fashion in science. It's a fashion that may help forge a new understanding of the complexity in the universe, the secrets of time and space, life and the brain, the mysteries of quantum physics, and the role of observers in reality.

A Demonic Paradox

I first began to appreciate the connections between information, quantum physics, and observers about a decade ago, after a conversation at an astrophysics meeting in Brighton, England, with Wojciech Zurek.

Zurek is a very energetic guy, a fast talker, full of ideas, and he tends to cram as many ideas as he can into every sentence. And he can string an almost endless stream of sentences together about nearly any scientific subject. Zurek is one of the great nonspecialists among modern physicists, or perhaps he would be better described as a multispecialist, because he does serious work in so many different areas. He turns up at astronomy meetings talking about galaxies, writes all sorts of papers about quantum theory, and organizes conferences on complexity and the physics of information.

Information and complexity are not really separate subjects from all of Zurek's other interests, though. Complexity is everywhere—in the mishmash of chemicals orchestrating thoughts and actions in the brain; in the summer thunderstorms swirling through the atmosphere; in the violent, hot, explosive whirlpools of cosmic matter streaming toward a black hole. Information is everywhere—in the ink forming symbols on paper, in the amounts of different chemicals in the bloodstream, in the way any object bigger than an electron is put together. In other words, Zurek is interested in everything. And for him, "everything" includes observers.

Observers, after all, unite information and complexity in the world. Complex systems are in essence information-processing mechanisms, systems that acquire information about the environment and use it to decide what to do next. Their acquisition and use of information, says Zurek, is what makes complex systems interesting—because acquiring and using information is the essence of being an "observer."

Humans, of course, are complex systems. And observers. But observers have traditionally not been welcome in physics. Physicists have long considered observers to be unnecessary complications, demons to be exorcised from the scientific description of reality. In other words, people are irrelevant spectators of the laws of nature in action. The universe should be the way it is whether anybody lives in it or not.

"Putting the observer into physics—physicists don't like doing things in this way," Zurek told me over lunch in a small restaurant across the street from the English Channel. "There were objective laws which were not being influenced by what the observer was doing. So physics was not supposed to have anything to do with the actual observation. . . . Observation was either shunned or ignored in various ways for a very long time, basically because people didn't know how to approach it."[3]

In particular, physicists did not really know how to approach a hypothetical observer proposed by one of the greatest of nineteenth-century physicists. The physicist was James Clerk Maxwell, and the observer was his demon.

Maxwell was the man who explained electricity and magnetism with mathematics after Michael Faraday had done so with words and pictures. Maxwell's equations remain today the cornerstone of the science of electromagnetism. But his demon entered the drama of physics on a different stage where Maxwell was also a player, in the story of thermodynamics.

Thermodynamics is a fancy word that scientists use to make the study of heat sound important. In their essence, the laws of thermodynamics merely describe the flow of heat and how heat is related to other forms of energy. In the nineteenth century, Sadi Carnot and his successors figured out that heat phenomena could basically be boiled down to two unbreakable laws. The first law is that the total amount of energy never changes—so whatever you do, you can't get more energy out of a device than you put in. The second law, expressed loosely, is that you can't even break even. More precisely, the second law requires any use of energy to produce some "waste heat" that is too degraded to do further work. In other words, nature tends to get messy. As "good" energy is transformed to waste heat, ordered structures maintained by energy tend to disintegrate. Order gives way to disorder. The technical term for such disorder is entropy; according to the second law, entropy always increases.

The first law of thermodynamics is widely known as the law of conservation of energy. The second law is widely known as the Second Law. When a physicist refers to "the second law," there is no need to ask which second law. Sir Arthur Eddington, the leading British astrophysicist of the first half of the twentieth century, called the Second Law supreme among the laws of nature. "If your theory is found to be against the second law of thermodynamics I can give you no hope; there is nothing for it but to collapse in deepest humiliation," Eddington wrote in the 1920s.[4]

But Maxwell's demon seemed to disagree with Eddington. The demon entered the world in Maxwell's 1871 book, Theory of Heat (although Maxwell had discussed it with colleagues a few years earlier). With "a being whose faculties are so sharpened that he can follow every molecule in its course," wrote Maxwell, the Second Law could be violated.[5]

Now you don't need to know the math to realize what shocking powers such a demon could have. It could perform feats of magic that would baffle David Copperfield. Melt some ice in a glass of hot tea, and the demon could return the cubes to their original cold, semicubic state. Scramble some eggs, and the demon could restore them to sunny-side up.

Of course, Maxwell did not assign his demon a task as ambitious as recooking breakfast. He merely wanted a little demonic heating and air conditioning. Picture two boxes of air connected by a gateway (or two rooms with a connecting door—the principle is the same). If the air temperature is the same in both rooms, the average speed of the air molecules will also be the same—that's what temperature means. The temperature of any gas depends on how fast its molecules are moving on average. Individual molecules, though, vary in speed, some jetting around at much higher than the average velocity and some much slower.

Imagine now that Maxwell's tiny demon can control the door between the rooms. As molecules approach the gateway, the demon could allow only fast ones to pass into one room, while permitting only slow ones to enter the other. After a while the fast room would be full of hot air; the slow room would be like a meat locker. A demon could therefore heat or air-condition a house without paying utility bills. That's better than breaking even. It's getting useful energy from waste heat, thereby breaking the Second Law.

Maxwell, however, never contended that the Second Law was fraudulent. He apparently meant to show that it was a "statistical" law—a law based on probabilities. Even if sorting the slow, cold molecules from the hot, fast ones could break the Second Law in principle, it wasn't possible in practice. Maxwell's point was that the Second Law depended on the insurmountable difficulty of tracing the paths of individual molecules in bulk matter.

Zurek thinks that Maxwell's demon showed something more than that—namely, that an observer does have something to do with physics. Observers bring additional information into the physicist's picture. The physicist's equations describing a gas provide "probabilistic" information about the temperature. You can precisely compute the average speed of the molecules, but not the exact speed of any single molecule—only the odds that it will be flying around at a given speed. But making an observation changes the situation. If you play the role of Maxwell's demon and look at any one of those molecules, and measure its speed, you get an exact (well, pretty much exact) answer. The trouble is there's no corresponding change in the physicist's description of the gas. It seems to Zurek that something has changed when the probabilistic information about a gas molecule becomes definite information. In other words, information is something real. Physics should take account of it.

As it turns out, taking information into account is the key to understanding why Maxwell's demon does not break the Second Law. The demon's innocence became clear only after physicists realized that the best way of posing the paradox was not in terms of energy, but of information.

The first step toward that realization was taken by Leo Szilard, a Hungarian physicist who coincidentally was also the first person to realize that the energy trapped in the atomic nucleus could be explosively released. Szilard was instrumental in the events leading to Einstein's 1939 letter to President Franklin Roosevelt urging the government to begin what became the Manhattan Project, which led to the development of the atomic bomb.

Years before, Szilard had spent a lot of time studying Maxwell's demon. And in 1929 he wrote a paper trying to settle the issue. A demon could not really break the Second Law, Szilard reasoned, because the gatekeeping job required finding out how fast an approaching molecule was flying. Making such a measurement—acquiring information about the molecule's speed—would doubtless require some energy. If the demon intended to beat the Second Law by extracting useful energy from a useless mix of molecules, it would have to make its measurements using less energy than it expected to get out in the end.[6] Alas, Szilard calculated, the demon would need more energy to make the measurements than it would get back in return for its efforts. On balance, useful energy would turn into disordered waste energy; entropy would increase. The Second Law could remain on the books.

Szilard did not, however, explain precisely where in the observing process the demon's energy bill would be paid. Szilard knew that information handling had something to do with the need for energy, but he did know exactly how—perhaps because in his day, before the era of the computer, physicists did not yet have a language for discussing information in a rigorous way. But that language of information was soon to be invented, in the form of a theory of communication that eventually came to be known as "information theory."

Shannon's Entropy

Information theory was born five decades ago and soon thereafter was presented to the world in the form of a book called *The Mathematical Theory of Communication*. The bulk of the book was a reprint of papers published by Claude Shannon in the *Bell System Technical Journal*. In those papers, which appeared in July and October of 1948, Shannon had shown how mathematics could be used to measure information, to analyze how much of it could be transmitted and how errors in information transmission could be reduced or eliminated. Since Shannon worked for Bell Labs, his findings were naturally useful in improving communication over telephone lines. But it was immediately obvious that his theory could be applied to any situation where information was involved.

Shannon's math measured information in bits. As Wheeler's

coin-tossing experiment illustrated, one bit is the information needed to decide between two possible messages for which the odds are equal. Specifying heads or tails takes one bit of information; specifying any one word from a list of two (such as "yes" and "no") also takes one bit of information. Specifying a single word from the entire dictionary, however, conveys a lot more information. A good unabridged dictionary contains something close to 500,000 words; designating one out of that many possibilities takes roughly 19 bits.[7] Shannon showed how to calculate how many bits are involved in any situation.[8]

Curiously, Shannon's equations looked suspiciously similar to the math behind the second law of thermodynamics. A key feature of the Second Law—the concept of disorder, or entropy—is described mathematically exactly the same way that Shannon described his way of calculating bits. This coincidence did not escape Shannon's attention. In fact, he named his new measure of information "entropy" to emphasize the similarity.

This raised the question: Is there a real connection between entropy in physics and the entropy of information? Is information theory merely convenient for calculations about sending signals, or does it possess some deeper significance? A few years ago I posed that question to Robert W. Lucky, who at the time was executive director of research at AT&T's Bell Laboratories. "Everybody who's looked at it and thought about it sees a great beauty in information theory," he responded. "Now the question is . . . are we deceived by the beauty into thinking that there are laws of nature that lie in there?" He thought the jury was still out.[9]

Shannon himself seems to have been similarly ambiguous on this question. "I think the connection between information theory and thermodynamics will hold up in the long run," he said in 1979, "but it has not been fully explored and understood. There is more there than we know at present."[10]

Many physicists take it for granted that information theory has something fundamental to say about reality. When Zurek helped organize a conference in 1989, he alluded to "a deep analogy between thermodynamic entropy and Shannon's information-theoretic entropy."[11] Now, a decade later, more and more scientists are saying that there is no deception in the analogy—laws of nature do indeed lurk in Shannon's equations. The physicist's entropy and Shannon's entropy are two sides of a coin.

Studies of the physics involved in information processing (especially with computers) have now solidified the conclusion that information is more than a metaphor. It is not something abstract and subjective, but as real as atoms, energy, or rocks. Information can be quantified. It is always embodied in some physical representation, whether ink on paper, holes in punch cards, magnetic patterns on floppy disks, or the arrangement of atoms in DNA. To me, this is one of the deepest and most important while least appreciated and least understood discoveries of modern science: Information is physical.

Landauer's Principle

The prophet of this point of view is Rolf Landauer, the IBM computer physicist widely known within the field of condensed-matter physics as one of its most perceptive critical thinkers. But he is not as well known outside his field as he deserves to be, for his work has contributed deeply to a new way of understanding physics in general. In essence, Landauer has done with computers what Sadi Carnot did with steam engines—extracted a new principle of nature from the workings of a machine. And as Carnot's work eventually led to the science of thermodynamics, Landauer's work is at the foundation of understanding the physics of computation.

Landauer came to the United States from Germany at age eleven, studied physics at Harvard, worked a short while for NASA's forerunner, and then joined IBM, where he became one of the world's foremost experts on the physics of computers and their components, like semiconductors. During the 1950s, he became intrigued with the notion of understanding the ultimate physical limits to the computing process. As he wrote years later, such studies had little to do with the technology of computer design. Real computers use vastly more energy than the fundamental lowest limit in principle.

"When reality is so far from the fundamental limits, the limits do not serve as a guide to the technologist," Landauer wrote. "Then why are we in this business? Because it is at the very core of science. Science, and physics most particularly, is expressed in terms of mathematics, i.e., in terms of rules for handling numbers. Information, numerical or otherwise, is not an abstraction, but is inevitably tied to a physical representation. . . . Thus, the handling of information is inevitably tied to the physical universe, its content and its laws."[12]

In his search for the ultimate limits of information handling, Landauer was guided by the examples of Shannon's search for the most efficient communication system and Carnot's quest for the ultimate in steam engine efficiency. "The steam engine as a model was very much on my mind," Landauer told me during one of my visits to IBM.[13] "The other model was Shannon's work." The point was to find a way of understanding fundamental limits on computing efficiency that didn't depend on any specific computing machine, the way Carnot's limit applied to any kind of steam engine.

"It's just an example of the fact that you can find limits . . . which are independent of the detailed technology that you use," Landauer said. "The common denominator between Shannon and the steam engine is that they find limits—in the case of Shannon you have to add a lot of clauses for the conditions in which it's valid, but they both find limits on what you can do which are independent of future inventions. . . . They don't depend on the detailed choice of technology. So we said, well, there must be a way of talking about the computational process and its energy limits without having to say we are talking about relay logic or vacuum tube logic—there must be a way of doing it on a more fundamental basis."[14]

The tradition of seeking a machine's ultimate limits of efficiency go back much further, more than a century before Carnot, to the exploits of a fellow Frenchman whose work illustrated the pitfalls of the enterprise. He's one of my favorite characters from the history of science. He was the French physicist Antoine Parent.

Idealized Efficiency

When a scientist dies, colleagues traditionally express their appreciation and affection with kind remarks and reflections. But sometimes it's hard to think of anything nice to say. The poor eulogizer for Antoine Parent could say only that Antoine was not as bad a guy as he appeared—he had "goodness without showing it."

Parent, it seemed, was rather tactless, highly critical of his scientific colleagues, and furthermore was not a very good writer. He was also a maverick, even by the standards of the seventeenth century. Born in 1666, he was educated as a lawyer but soon turned to math and then physics, perhaps indicative of the rebellious streak that won him few friends in the French scientific world. He wrote on a wide range of scientific problems, often stressing the flaws in Descartes' scientific system, which was also no way to win favor from the French establishment, for whom Descartes was the cherished authority.

Parent had some admirable qualities, though. One was his recurring concern with applying scientific reasoning to matters of practical importance. One such practical matter that seized his attention was the issue of the ultimate possible efficiency in the operation of industrial machinery. As a model for studying this question, he took the waterwheel.

In seventeenth-century Europe, water was the most important source of force. Waterwheels had long driven mills for a variety of purposes—mostly grinding grain, but also for operating pumps and wood saws. While waterwheels were not exactly good machines to serve as models of the entire universe, they were important industrial tools. Because of their economic importance they warranted scientific study.

Parent studied a particular scientific question about waterwheels—their potential efficiency. In other words, how much of the energy in flowing water could an ideal waterwheel extract? The answer must depend on how fast the waterwheel turns—but not, as the historian of technology Donald Cardwell has pointed out, in an obvious way.[15] A wheel connected to too heavy a load would not turn at all and could therefore do no work. If the wheel was connected to nothing, it would turn as fast as the water flowed. But it would be doing nothing but spinning. Again, no work. So the most work had to correspond to some intermediate spinning speed.

As it turns out, the answer is deceptively simple. The most work is done if the wheel spins at one half the speed of the stream, and this maximum efficiency mode allows the wheel to capture half the stream's energy. But this answer was in no way obvious at the beginning of the eighteenth century, when Parent took up the problem. He considered the case of a frictionless wheel with blades struck on its bottom side by water flowing in a stream. The impact of the water on the blades imparted motion to the wheel.[16]

Parent reasoned that the force of impact from the water depended on how much water hit the wheel's blades and on how much faster the stream flowed than the wheel turned. Using some relatively advanced math for his day, he concluded that at best a waterwheel could convert only about 15 percent (4/27ths) of a stream's energy into useful work. In other words, four twenty-sevenths of the water could be raised back to its original height by a perfect waterwheel driving a perfect pump. The waterwheel's speed at maximum efficiency, Parent calculated, would be one-third the speed of the stream.

This would have been a great achievement, and Parent would be famous today instead of unknown, except for one thing. He was wrong. As Cardwell observed, Parent neglected to idealize the wheel, assuming that some water would pass without hitting the blades. An ideal wheel would have blades close enough together to prevent that loss of energy. So as a matter of principle, Parent badly underestimated how efficient a waterwheel could be. Nevertheless, Parent's efforts were significant as a pioneering example of the scientific study of machine efficiency. Cardwell noted that "Parent's paper . . . inaugurated research into the efficient exploitation of energy resources, research whose importance today need not be stressed. And it introduced—if only by implication—a fundamental concept for both science and technology. His yardstick was the extent to which the initial situation could be recovered. . . . This criterion of 'reversibility' was, very much later, to become a basic concept in the nineteenth-century science and technology of thermodynamics."[17]

To me, the interesting thing about the Antoine Parent saga is how it foreshadows the modern study of the efficiency of computing. Just as Parent underestimated how efficient a waterwheel can be—by failing to analyze the ideal case—modern physicists have long underestimated how efficient in principle a computer can be, also by failing to appreciate the ideal case. Landauer led the way to new understanding of the true situation, which could be described in much the same words as a computer maker's ad campaign: "computing without limits."

Landauer had never heard of Antoine Parent. His main sources of inspiration were Carnot and Shannon. But there was also inspiration of a different sort from Leon Brillouin, whom Landauer had encountered while at Harvard. In the mid-1950s, Brillouin's book *Science and Information Theory* had attempted to survey the principles of physics and information in a way that Landauer found unsatisfying. "The book left me with the reaction: 'There must be a better way to think about that.' "[18]

In his book, Brillouin attacked Maxwell's demon with the assumption that the demon couldn't see in the dark. Therefore it would need to shine a light on an approaching molecule in order to see it and measure its velocity. The energy needed for the light would exceed the energy the demon saved, preserving the Second Law. Other scientists of the time explained away the demon paradox in a similar way.

Ultimately this approach was based on Shannon's calculation of how much energy it costs to send a message. Measuring something means acquiring information, which is basically the same thing as communicating. And Shannon showed that sending a bit of information over a telephone line required a minimum amount of energy, roughly the amount of energy possessed by a typical molecule bouncing around at room temperature.

Landauer saw into the situation more deeply than Brillouin and others. Landauer realized that if measuring is like communicating, it is also like computing. Measuring, communicating, and computing are all about exchanging information. If Brillouin was right, then computing would require an unavoidable energy loss at every computational step, the same as every molecular measurement by Maxwell's demon. But Landauer saw that there is more than one way to communicate. Shannon's analysis was for a special case. There are ways to send messages other than by waves over wires.

"I don't have to communicate in that way," Landauer liked to say. "I could hand you a floppy disk. If I'm desperate, I could use the U.S. Postal Service."

So in the late 1950s, Landauer took a fresh look at computing physics, first in collaboration with John Swanson, and then on his own after Swanson's accidental death. Landauer's landmark paper arrived in 1961, published in the *IBM Research Journal*. There he explained that computing in and of itself did not require any minimum use of energy. Bringing two systems together to allow information exchange can be done in ways that do not lose any energy. An ideal computing system could in principle manipulate information while creating no waste heat at all. The key was in realizing that a computation could in principle be carried out more and more slowly to reduce any friction that would generate waste heat. It was just as with Parent's waterwheel, where in principle you could always put the blades closer and closer together to reduce the amount of wasted water. In principle you could always compute more slowly if you wanted to reduce friction and waste less energy, all the way down to zero energy loss.

Except for one hitch. Landauer believed in 1961 that a computer could not really work without dissipating (that is, using and losing) energy. A real computer does more than just compute. It also erases information in the process. Ordinarily, when a computer multiplies two numbers together and gets an answer of 54, there is no need to remember whether the original numbers were 6 and 9, 18 and 3, or 27 and 2. Only the answer matters; the computer "forgets" the input. It is in the forgetting, or erasing, of information that computers suffer an unavoidable loss of energy. Erasing a single bit, Landauer calculated, requires an energy loss roughly equal to the energy possessed by a bouncing molecule. In other words, erasing information always produces heat that escapes into the environment.

Proving this requires some high-level mathematics that you shouldn't attempt to do at home. You can, however, perform a home experiment to illustrate the principle if you have a couple of basketballs.

The two basketballs provide a system for representing information. Put one on the floor by your left foot and hold the other in your right hand. A ball on the floor represents 0; a ball in your hand represents 1. You are now playing the role of a human floppy disk with an information storage capacity of two bits.

Now, your task is to erase the bit (1) in your right hand. Sounds easy enough—just drop the ball. But the ball does not then simply occupy the 0 state, on the floor. Instead, it bounces. In fact, if you have a perfectly elastic basketball, and a good hard floor, it will bounce right back up to your hand, the 1 position. To settle down into the 0 position—to erase the 1 bit—the basketball has to encounter friction, with the air molecules and the floor. So after each bounce the ball will rise to a lower level. Eventually the friction slows down the ball, it stays on the floor, and the 1 has been erased. But why? Only because energy from the bouncing ball has been transmitted to the floor and the air. In a vacuum with a frictionless floor, the ball *would* bounce back up to your hand every time. The information can be erased only if energy is used up in the process.

Landauer's great insight was that this energy loss is necessary no matter how you erase information, regardless of what kind of information it is. Think about it. If you erase a pencil mark with an ordinary eraser, you generate all kinds of heat energy through the friction between the eraser and the paper. Burning a book (an ugly thought) nevertheless illustrates the energy requirement to eliminate information. The fact that erasing a bit requires a minimum loss of energy is now widely known among computer physicists as Landauer's principle.

Landauer did not apply his principle to Maxwell's demon. But years later, his colleague Charles Bennett at IBM realized that Landauer's principle applied to demons just as much as to basketball players. Landauer had not gone far enough in his analysis. Suppose a computer did not throw away any information. It could then avoid any loss of energy. True, keeping track of the intermediate results of every calculation would require the mother of all hard drives—but seemed thinkable in principle. With enough gigabytes, maybe Maxwell's demon could beat the Second Law.

But of course, this trick also fails. Somebody has to build the hard drives. An accurate account of the manufacturer's utility bills would show that the demon's energy savings would be offset by the amount needed to make the hard drive, or magnetic tape, or whatever other form of energy storage the demon adopted. So the Second Law survived, and Maxwell's demon was prepared for burial. Nobody, demonic or otherwise, could get good energy for free.

Reversible Computing

Charles Bennett, of course, was not trying to get energy for free. He worked for IBM, which in those days could afford to pay its utility bills. But since it was IBM, Bennett was interested in new ways of computing. And in analyzing the energy needed to compute, Bennett came to a stunning conclusion—maybe you couldn't get energy for free, but maybe you could compute without using up energy.

Landauer had shown that you needed energy to compute because you had to erase information along the way. Throwing away information requires loss of energy. You cannot get around it. But wait, said Bennett. Suppose you figure out how to compute without erasing any information. Then you wouldn't have to lose any energy.

At first glance, such a computational scheme would run into the same limits that afflicted Maxwell's demon. Sure, you could compute without losing information if you stored in memory all of the intermediate steps of your calculations, never throwing any information away. But you would have to shell out money to buy bigger hard drives or a bunch of Zip disks, all requiring energy to build. Even if you didn't worry about that energy, your computer would get bigger and bigger as you added storage space. Sooner or later you'd run out of room for all the equipment.

But suppose you didn't try to save all the intermediate information, Bennett asked himself. If all the steps in the computing process could be reversed, he reasoned, then any previous information could be reconstructed. Suppose that you designed the computational steps so that they could be retraced, without committing each individual step to memory. You would program the computer not to remember everywhere it had been, but to figure out how to go back to where it had been. In other words, at any point in a computational stream you could throw the gears into reverse and send the computer back to its starting point. Such a "reversible computer" could retrace all the logical steps involved in performing its computations. Run the program forward to get your answer; run it backward to recover all the information that was involved in getting that answer.

Bennett worked out how to design the hardware for reversible computing in 1973. His paper showed that a computer, in principle, doesn't really have to erase any information. Therefore computing could be done expending as little energy as desired, without the need for massive memory storage. "If you do it just right, it doesn't slow you down very much or make you use very much memory," Bennett said.[19] It's all a matter of setting things up so that the computer can undo what it has done. Think of crossing a river using a set of magic stepping stones, Bennett advises. You must pick the stones up and place them back on the water according to certain rules. Observing those rules, you can get across the river with only a few stones. Yet knowing the rules also allows you to retrace your steps precisely, so the information about how you crossed the river is preserved. In the same way, reversible computers—using the right rules—can work out complex computations by doing and undoing things in a way that preserves information about every step along the way.

When Bennett first proposed reversible computing, Landauer himself was dubious. For months, he suspected that something was amiss. Even Richard Feynman, perhaps the greatest genius in physics of his day, was skeptical initially.[20] Landauer eventually realized that Bennett's idea worked in principle. But even today many other scientists don't understand or appreciate the impact of Bennett's discovery of reversible computing or his explanation of Maxwell's demon. So it's important to make the fundamental point clear: computing does not require any loss of energy. In handling information, the only unavoidable loss of energy is in erasing it.

Believe me, this is right. But you will find other sources that say otherwise. One recent popular book, by a very capable scientist and writer, says flat out that "there is an absolute minimum amount of energy required to process a single bit of information." That simply isn't true, as Landauer has emphasized on many occasions. "There really aren't any limits to information handling," he insists. "Measurement does not require energy dissipation. Neither does communication."

In principle, at least. But whether energyless computing will ever make anybody any money is another issue. For many years Landauer felt that Bennett's reversible computing scheme was of interest only for fundamental physics, not for making real computer chips. For two decades it was a curious bit of computer theory lore known only to a handful of computer physicists[21] and careful readers of Scientific American. But in the 1990s, reversible computing came to life.

Ralph Merkle, a computer scientist with the Xerox research center in California, saw that reversible computing didn't have to be perfect to be practical. Sure, building a completely reversible computer was an unrealistic goal. But you could still save some energy even if only some of the computing was reversible.

Merkle unveiled his idea at the 1992 physics of computation meeting in Dallas. It didn't get much attention, though, until a few months later at the American Physical Society's March 1993 meeting in Seattle, the meeting where Bennett introduced the world to quantum teleportation. "You can go to Radio Shack, buy the right parts, solder them together and have some sort of . . . reversible gadget, if you do it right," Merkle said there. "Reversible logic," he asserted, "will dominate computing in the twenty-first century."[22]

The reason for that, he argued, is that computers are energy hogs. Between 5 and 10 percent of the electrical power produced in the country feeds computers. Reversible computing could therefore ease some of the drain on the nation's energy supply. Apart from such altruistic motivation, reversible computing could help the consumer, too—by extending laptop battery life, perhaps, or helping to prevent smaller, faster computer chips from overheating. Chip designers always want to condense more circuitry into smaller spaces so electronic signals have shorter distances to travel, allowing the chips to work faster. But confining a lot of computing to a small space causes excessive heat to build up from the energy used in the computing process.

In general, every computing step, or "logic operation," uses up a certain amount of energy that gets discarded as heat. Over the years, computer designers have managed to steadily reduce the amount of energy used per step. But that trend can't continue forever. Eventually they will run into Landauer's principle. When the energy per logic operation nears the average energy of a single molecule bouncing around at room temperature, the need to erase intermediate steps prevents any further energy savings.

Bennett's reversible computing scheme avoids the limit imposed by Landauer's principle. So the goal of energy-conscious computer designers, Merkle argued, should be to use the kinds of reversible steps that preserve the initial information as much as possible. "Some instructions in a computer are reversible and other instructions are irreversible," Merkle said. You don't have to have complete reversibility for the whole computation. Clever design of the hardware can make most of the steps reversible, saving a lot of information and therefore saving a lot of energy.

Landauer pointed out that it is too soon to say whether even such limited reversible computing schemes will really turn out to be practical. For one thing, they require special kinds of power supplies capable of extracting the electrical energy out of the circuitry when information is recovered. But the essential fundamental point remains that information is real and physical. Reversible computing illustrates how important it is to take the reality of information into account when designing computers.

Until the mid-1990s, though, information's role in physics was still mostly what outsiders call "inside baseball"—curious technicalities pondered by specialists but of little interest to the spectators outside the field. But lately scientists in other areas—and popular media—have begun to take the reality of information more seriously, thanks to the prospect of a machine potentially more powerful than even Edward Fredkin imagined—the quantum computer.

Chapter 4

The Quantum and
the Computer

Classical and quantum information are very different. . . .
Together, the two kinds of information can perform feats
that neither could achieve alone. . . . These concepts could
result in a revolution in computer science that may dwarf
that created decades ago by the transistor.

—GILLES BRASSARD,
"New Trends in Quantum Computing"

If there's a visitor from the future among us today, it must be Robert
Redford.

Think about it. *Three Days of the Condor* (Robert Redford, Faye
Dunaway, Paramount Pictures, 1975)—all about a secret plan to
fight a war in the Middle East over oil. They should have called the
movie *Operation Desert Storm*.

And then there was *Sneakers* (Universal Pictures, 1992). Somehow that film foresaw the computer of the future. It was a computer
unlike any other, a computer capable of calculations undreamed of by
computer science's pioneers. It was—although nobody said so in the
movie—a quantum computer.

In *Sneakers*, Redford's computer-hacker security consultants get their hands on this computer—hidden in a black and gray box disguised as a Panasonic phone answering machine—without really knowing what it is. Inside the box they find a computer chip designed by a mathematician named Gunter Janek (played by Donal Logue). Then the blind hacker on Redford's team (David Strathairn) connects the box via modem to the Federal Reserve, the national power grid, and the air traffic control system. All the encoded secrets at every site are instantly decoded. Somehow, Janek's chip can crack the toughest cryptography systems in the world.

"Cryptography systems are based on mathematical problems so complex that they cannot be solved without a key," Strathairn explains. Evidently Janek figured out a program for solving those problems and wired it into the chip.

"So it's a code breaker," says River Phoenix, another member of Redford's team.

"No," says Redford. "It's *the* code breaker."

Redford's hackers couldn't explain the details behind the magic box's code-breaking ability. After all, this was fiction. What *Sneakers* envisioned in 1992 was not really possible then in the real world. No one had any idea how to design *the* code breaker. But two years later, film fiction became physics possibility. A Bell Labs mathematician figured out how the toughest codes known could have been cracked, if the mysterious box in *Sneakers* contained a quantum computer.

Sneakers opened in U.S. theaters on September 11, 1992. Less than a month later, most of the world's quantum computer aficionados met in Dallas, at the conference where Benjamin Schumacher introduced the idea of qubits to measure quantum information. I don't recall anyone mentioning *Sneakers* at this meeting, though—back then, nobody knew that quantum computers could be used to break codes. Several speakers at the meeting did discuss quantum computing, however, describing how the multiple possibilities of the quantum realm could be exploited to conduct complex computations.

Some in the audience didn't take the idea very seriously. Throughout the meeting I sat next to Rolf Landauer, the world's leading skeptic on quantum computing's long-term commercial prospects. At one point I recall his objecting that a speaker was describing quantum computers as though they could really be built. I forget the speaker's response, but I remember Landauer's reply—to the effect that he was still waiting to see somebody file a patent disclosure.

Landauer had several good reasons for suspecting that quantum computers were being oversold. For one thing, he was right that nobody really knew how to build a quantum computer. Second, there was no important problem that anybody could think of that you'd need a quantum computer to solve. Third, even if there was such a problem, you would not be guaranteed of getting an answer. If a quantum computer worked three times as fast as an ordinary computer, you would get the answer you wanted only one-third of the time.

Research in quantum computing, therefore, was not really inspired by the search for a way to get rich by putting IBM out of business. It was a matter of using quantum computing to understand more about the physical nature of information—how to describe the world, in principle, in the language of information processing. Even if you couldn't really build a quantum computer, you could imagine how one might work, and that might in turn give insights into the nature of reality. After all, if reality is quantumlike at its foundation, and you want to describe reality in terms of information processing, you have no choice but to imagine quantum information processing—or quantum computing.

Popular accounts sometimes credit the origin of this idea to Richard Feynman, the famous physicist who gained public notoriety as a member of the presidential commission investigating the *Challenger* explosion. But Feynman was not actually the first to imagine a quantum computer. Paul Benioff, a physicist at Argonne National Laboratory, explored the idea as early as 1980. He wondered whether a quantum process could perform computations (in other words, simulate a Turing machine) and found the answer to be yes. He didn't investigate whether quantum computing could do anything special or more powerful than an ordinary computer.

Feynman viewed quantum computing in broader terms. Of course, Feynman viewed everything in broader terms. And in different terms. Somehow, Feynman seemed to see the world differently from anybody else. As the subtitle of his autobiographical reminiscences (*"Surely You're Joking, Mr. Feynman!" Adventures of a Curious Character*) suggested, he was truly a curious character. If Redford is from the future, Feynman was from another planet.

Feynman's legendary exploits have already filled several books, some by Feynman and friends and other historical accounts and biographies. He was far more than a physicist. He was also an artist and musician—the greatest bongo drum player science has ever known. Before a trip to Japan, he taught himself Japanese. On his honeymoon he became an expert in Mayan astronomy. While working on the atomic bomb project during World War II, he learned to pick locks and crack safes. He was rejected by the army because psychiatrists thought he was crazy. He was awarded the Nobel Prize because his craziest ideas turned out to be correct.

Among the physicists of the mid-twentieth century, Feynman stood out not only by personality, but by his genius. I once asked Hans Bethe, himself one of the century's most respected physicists, about Feynman. They had worked on the Manhattan Project together and had been colleagues at Cornell before Feynman moved to Caltech. When I mentioned Feynman, Bethe's eyes lit up.

"He was a phenomenon," Bethe said. There were many geniuses among physicists, but with most of them you could figure out how they did what they did. Feynman, on the other hand, "was a magician."

"I could not imagine how he got his ideas," Bethe said. "Feynman certainly was the most original physicist I have seen in my life, and I have seen lots of them."

I then asked Bethe—who knew all the great physicists of the century—if there were any other magicians. "No," he replied. "Just Feynman."[1]

Consequently, when Feynman talked, other physicists listened. And toward the end of his life, he had turned his attention toward the physics of computing. At the first major meeting to consider the physics of computing, in 1981, he gave a major address on the issue of how closely computers could, in principle, simulate reality.

An ordinary computer could never simulate nature in all its details, Feynman pointed out. An ordinary computer worked deterministically, while nature exhibited all those mysterious quantum properties that could not be simulated so simply. While Benioff had correctly shown that a quantum process could simulate ordinary computation, Feynman noted that it was a one-way street—ordinary computation could not simulate a quantum process.

Feynman's discussion naturally led to a consideration of the Church-Turing thesis and the limits of Turing machines to compute (or "decide") certain questions. There is a deep question that the thesis in itself does not answer. Do undecidable questions impose a limit only on human (or machine) abilities, or is reality itself restricted by the same limit? In the context of ordinary computing, Feynman realized, this wasn't even the right way to frame the question. Computers back in 1981—any universal Turing machine—operated strictly according to the laws of classical mechanics. A classical computer cannot simulate "real" physics because reality is quantumlike. So you can conclude nothing about limits on reality from the limits of a classical computer.

"Nature isn't classical, dammit," Feynman said, "and if you want to make a simulation of nature, you'd better make it quantum mechanical." Trying to do so, he said, was "a wonderful problem" because "by golly . . . it doesn't look so easy."[2]

Perhaps it's not obvious to everybody why ordinary computers can't simulate a quantum universe. After all, computers simulate a lot of complicated things—like baseball games and the world economy. But remember, the quantum realm encompasses all those multiple possibilities. Simulating a single baseball game is easy compared to simulating all the baseball games that have ever been played (all at the same time). And that would be only a tiny fraction of all the possible baseball games. Besides, when you're talking about simulating reality itself, you're talking about all the different possibilities for all the atoms in all those baseball players. No ordinary computer, no matter how powerful, can compute all the quantum possibilities for a large number of particles. You'd need a computer that can reproduce the multiple-possibility world that quantum physics describes. It would take a computer built of quantum mechanical elements that could preserve all of those many quantum mechanical possibilities.

Such a computer, Feynman conjectured, might be able to replicate natural processes precisely. It would then be possible to make calculations about the behavior of molecules, atoms, and smaller particles that only quantum math can accurately describe. Such a quantum computer could cope with all these multiple possible realities that quantum math contains.

Feynman could find no reason why such a quantum simulation of nature shouldn't be possible. "It turns out, as far as I can tell, that you can simulate this with a quantum system, with quantum computer elements," he said. "It's not a Turing machine, but a machine of a different kind."[3] He did not, however, provide any instructions for building such a machine.[4]

A few years after Feynman's talk, another physicist saw that quantum computing could do more than just simulate natural processes. David Deutsch, at the University of Oxford in England, realized that a quantum computer could compute much more rapidly than an ordinary computer. Deutsch was an advocate of a particular view of quantum physics called the many-worlds interpretation. (See chapter 9.) He believed (and still does) that multiple universes coexist, with the different quantum possibilities all becoming real in some universe or another. It's as if the universe is a chess game, and all the possible games are being played all at the same time. At first the various games are all alike, but as the games progress, different moves create different arrangements of the pieces on the board.

Now if you're playing Chessmaster 4000 on a Pentium-based PC, you can see only one arrangement of pieces at a time. But with the quantum Pentiums of the future, you might see a board where a fuzzy bishop occupied several different squares at once as your quantum processor preserved all the possibilities for all the moves up to that point in the game.

If the many-worlds view of quantum reality is correct, Deutsch argued in 1985, then a properly designed quantum computer could compute in all those universes at once. In principle, such a quantum computer could perform millions upon millions of computations while your Tandy 1000 (a hot computer in those days) could work on only one.

Deutsch's proposal amounted to a strange new kind of parallel processing. The idea of multiple processors running simultaneously (in parallel) was not new. Parallel processing has long been a big deal among computer designers. Obviously, 1,000 chips running in tandem could solve problems faster than one chip making one calculation after another. Basically, a thousand chips could get your answer roughly 1,000 times faster (depending, of course, on how well the machine was designed and programmed).

But quantum parallelism was vastly more impressive. Consider the power of 1,000 simple processors. Say each can represent one bit of information (in other words, be on or off, representing 1 or 0). There are then 2 to the 1,000th power possible combinations of on-off sequences. (This is a big number, roughly 1 followed by 301 zeros. If you tried to count up to it—at the rate of a trillion numbers per second—it would take more than 10 billion times the age of the universe. Don't even try.)

Obviously, no ordinary digital computer could ever sort through all the possible combinations of that many processors; in other words, there would be some problems that such a computer simply would never have enough time to solve. But if you replace the standard Intel versions with quantum processors—all recording qubits, or mixtures of 0s and 1s—then all the possible combinations of 0 and 1 exist at once. In essence, this translates into a computer with a speed 2 to the 1000th power times faster than the speed of a single processor.

Put another way, suppose you had to open a lock with one of 1,000 keys. Trying each of the 1,000 keys would take quite a while. But if it's a quantum lock-and-key, you could try all the keys at once. Or you could think in terms of the code-breaking problem in *Sneakers,* breaking big numbers into prime factors. Code makers always believed they could defeat any code-breaking computer simply by making the key number longer. Nowadays, hundreds of computers working together over the Internet need months to crack numbers with 150 digits. At that rate, factoring a 1,000-digit number would take more than a trillion trillion years. But a quantum computer, in theory, could do that job in twenty minutes.

So why, back in 1985, wasn't Bill Gates designing MS-QOS instead of updating DOS? Because for all the hypothetical power of a quantum computer, there was no problem anybody could think of to use it for. The reason goes back to the fragile nature of quantum information. Sure, a quantum computer could work on a problem in countless universes at once (the way Deutsch looked at it).[5] But a computer working on a problem in (say) a trillion universes would produce a trillion answers. And there was no way to know which was the answer you wanted. If you looked at one answer, you destroyed all the rest. You would have to be extraordinarily lucky for the answer you looked at to be the one you wanted.

So if you designed a quantum computer to work ten times faster than an ordinary computer, it would find the answer in one-tenth the time. But you would have only a one in ten chance of getting the answer you wanted when you looked. A wash.

That was roughly the way things stood when *Sneakers* appeared in 1992. By then Deutsch and his colleague Richard Jozsa had in fact proposed a problem that a quantum computer could actually solve faster than an ordinary computer, but it was not a problem that anybody would pay real money to solve. It was a "toy" problem, designed merely to illustrate the principle of quantum computing. It had to do with deciding which of two possible mathematical operations were being performed on a number. Roughly, suppose you wanted to know whether Bob the mathematician was a devotee of formula 1 or formula 2. You send him numbers from a long list one at a time; he performs a calculation on the number, using his favorite formula, and then sends you the answer. You then send him another number, and the process is repeated. The question is, how many times must you go through the process to figure out which formula Bob is using? Using ordinary computers, you might have to send more than half the numbers on your list—if you started with 99 numbers on the list, for example, you might have to send 51 of them before figuring out which formula Bob was using. With quantum computers, you would need to send only two.

The Deutsch-Jozsa problem illustrated the power of quantum computing, but not its value. Nobody really cared which formula was Bob's favorite. But in the next two years, the premise of *Sneakers* came alive, connecting quantum computing with the world of espionage and international intrigue.

First came advances in the math underlying quantum computing from Umesh Vazirani and his student Ethan Bernstein at the University of California, Berkeley, and Daniel Simon at the University of Montreal. Then in 1994, building on that work, mathematician Peter Shor at Bell Labs saw a way to accomplish precisely the code-breaking power contained in the magic box in *Sneakers*.

As you probably don't recall from chapter 1, the most popular strategy in modern cryptography is to make a key for a code freely available. The catch is that these keys work only one way. They enable you to code a message, but not to decode one. The key is based on a very long number, more than 100 digits long. If you know the long number, you can encode messages. But to decode them you need to know the prime numbers that, when multiplied together, produce the long number. Figuring out the primes is the "hard problem" that David Strathairn alluded to in *Sneakers*.

It's eerie to watch *Sneakers* now and realize how close it came to forecasting Shor's discovery.

"There exists an intriguing possibility for a far more elegant approach" to factoring big numbers, Gunter Janek says in a lecture attended by Redford and his ex-girlfriend.

"This isn't just about large number theory, it's about cryptography," the ex-girlfriend comments.

"It would be a breakthrough of Gaussian proportions and allow us to acquire the solution in a dramatically more efficient manner," Janek continues. "I should emphasize that such an approach is purely theoretical. So far no one has been able to accomplish such constructions—yet."

Just a year and a half later, that "more elegant approach" to factoring appeared. Just as Janek in the movie had figured out how to solve those tough problems that protect the keys for secret codes, so had Peter Shor. Shor found his solution in March 1994. By mid-April, his breakthrough was circulating around the Internet in the form of a paper called "Algorithms for Quantum Computation: Discrete Log and Factoring." "Currently, nobody knows how to build a quantum computer," Shor wrote in that famous paper. "It is hoped that this paper will stimulate research on whether it is feasible to actually construct one."

It did.

A month after Shor's paper hit the Internet, I went to New Mexico for the Santa Fe Institute's workshop on the physics of information, where I learned about the developments in quantum information theory I described in chapter 1. The organizers realized that Shor's new algorithm was the hottest topic in computing and tried at the last minute to get him to come. He was unavailable, though, so Umesh Vazirani presented a talk on Shor's result. Vazirani offered his own analogy to illustrate the drama of Shor's achievement. Consider the problem of factoring a number 2,000 digits long, Vazirani said.

"It's not just a case that all the computers in the world today would be unable to factor that number," he pointed out. "It's really much more dramatic. . . . Even if you imagine that every particle in the known universe was a computer and was computing at full speed for the entire known life of the universe, that would be insufficient time to factor that number."

Shor's method, on the other hand, could break such a number down quickly. (At the time, it wasn't clear how quickly—maybe hours, maybe days, but certainly in a "reasonable" amount of time compared to classical computers.) Using Shor's approach, a quantum computer could find exactly what a mathematician needed to plug into a recipe for finding prime factors.[6] "This is the first really useful problem that has been shown to be solvable on a quantum computer," Vazirani said at Santa Fe. "It's a truly dramatic result."

All of a sudden, quantum computing became a subject worthy of research grant proposals. Factoring big numbers was a big deal. Spies, the military, the government, banks, and other financial institutions (and as *Sneakers* suggested, organized crime) all depended on systems offering super-secure privacy. Here was a way, in principle, that the best security could be defeated. And if one such use for quantum computers was possible, perhaps there would be others.

Within a year or two, concerns about quantum computing spread throughout the military-industrial complex like a computer virus on the Internet. The U.S. National Security Agency and other defense organizations, skilled at deciphering foreign codes, started scrambling to decipher the latest quantum physics research. By 1998, the NSA was shelling out about $4 million a year on quantum computing research conducted at various universities. The government's Defense Advanced Research Projects Agency, known as DARPA, was spending about $3 million. Universities and organizations ranging from IBM to NASA had all jumped into the quantum computing sweepstakes.

All that attention began to turn up further possible uses for a working quantum computer. Researchers suggested that quantum computers might provide speedier searches of large databases, better gyroscopes for guided missiles, and faster methods of working out airline schedules. Ironically, experts showed how quantum computers might prove essential in sending new secret codes that even quantum computers could not break. Quantum computers might also be able to simulate many natural processes too complicated to calculate, such as what happens when particles collide in the core of a hydrogen bomb blast, or how atoms stick together in more temperate chemical reactions (along the lines of Feynman's original idea for simulating reality on the molecular level). Quantum computers could shed light on the earliest moments of the universe and the innermost workings of organic molecules—all the sorts of things that encompass too many variables for any classical computer to handle.

Still, the larger question looming behind all the hype and hoopla was the one that troubled Landauer back in 1992. Can you really build a quantum computer that works?

I recall during the Santa Fe meeting that Landauer spent the long afternoon breaks after lunch trying to explain to Vazirani and others why building a quantum computer would not be as simple as it seems on paper. Ordinary computers, Landauer stressed, work only because they don't have to be perfect. Unless everything works perfectly, a quantum computer won't work at all. "And nature abhors perfection," Landauer said.

For one thing, to perform many simultaneous computations, a quantum computer requires almost total isolation from the environment. Any observation or interaction with the outside world of the slightest kind would eradicate the multiple computations. For another thing, Landauer argued, imperfections in manufacturing will cause a quantum computer to deviate from the precise processes envisioned by the quantum mathematics. These defects will likely abort the computation before it is completed. So on the one hand, a quantum computer must be designed so that its elements are very unlikely to interact with anything. Yet to compute, the elements must interact with each other. It would be difficult to build a device that accommodated both requirements.

The biggest problem, in Landauer's mind, was the need for precision far exceeding anything that ordinary computers ever accomplish. In today's computers, 1s and 0s can be represented by voltages in a circuit that do not have to be controlled to infinite precision. A low voltage means 0, a substantially higher voltage means 1. If the low voltage is off a little from specifications, that's okay—it's still obviously supposed to be a 0, and the machine can take steps to correct the small deviation from the designed voltage level. You can think of it like opening and closing a door, Landauer said. To close a door you just need to push it hard enough so that it will latch—you don't need to push with a precisely controlled amount of force.

But quantum computing elements are inherently different. Pretty close to zero is not the same as zero—it represents a mixture of 0 and 1 with a higher probability of zero showing up when a measurement is made. You can't "correct" that without throwing away the very information that makes a quantum computer more powerful.

Throwing away information introduces another problem that Landauer knew well—it costs energy. Quantum information is fragile, and for a quantum computer to work, its computing elements need to be isolated from the environment. Interacting with the outside word is a surefire way to destroy the quantum information. Erasure requires just such an interaction, because it can't be done without dumping some waste heat outside the computer. Even without erasure, maintaining the kind of isolation a quantum computer requires is hard to imagine.

Nevertheless, a lot of people began to imagine it.

Less than two years after the appearance of Shor's algorithm, quantum computing had become hot enough to warrant a session at science's equivalent of the World Series, the annual meeting of the American Association for the Advancement of Science (AAAS).[7] It was February 1996 in Baltimore, and by then several proposals had been made for how to build the computing elements—the logic gates—that a quantum computer might be made of. Two of those ideas had actually been demonstrated in the laboratory. Shor and others had made progress in figuring out how to correct errors in a quantum computer without losing the important information, and he came to Baltimore to talk about those error-fixing procedures. Seth Lloyd of MIT presented a proof of Feynman's original conjecture that a quantum computer could simulate any quantum system. Landauer was also there, pointing out once again that the barriers to success were still formidable and likely insurmountable.

The key advance by that time was the demonstration of quantum logic gates. Logic gates are the building blocks of all ordinary computers. Each logic gate consists of a group of transistors arranged to take in electrical signals representing 0 or 1 (the input) and then produce an output signal based on what comes in.

It's easy to diagram on paper how different kinds of gates can perform various logical operations. An AND gate, for example, takes in two input signals and checks to see if they are both 1s. If so, the gate spits out a 1 as well, signifying that the AND logic condition has been satisfied. Otherwise, the AND gate sends out a 0. An OR gate, on the other hand, would send out a 1 if either of its two input signals was a 1. A NOT gate flips any input signal to the opposite number: If a 0 enters, a 1 comes out, and vice versa. A NAND gate spits out a 1 if the inputs are both 0, and a 0 if the inputs are both 1s.

In quantum physics, of course, logic is more complicated. The qubits of quantum information carry mixtures of 0 and 1. A qubit is therefore something like a spinning roulette wheel; when it stops it will turn up red or black, but not necessarily with equal odds. In a quantum logic gate, qubits interact to produce an output with different odds of red or black—a different mixture of 0s and 1s. Because of the quantum world's peculiarities, quantum logic gates are a little more intricate than those of classical computers and have to play by slightly different rules (for one thing, quantum logic gates need to have the same number of inputs as outputs). The important point is that if you can make and hook up the right kind of quantum logic gates, you can do any quantum computation you like. In fact, as a team of quantum computation experts showed by 1995, all you really need is a "controlled-NOT" gate, and with the right arrangement, you can perform any of the other logic operations.

The first papers describing quantum logic gates in the lab appeared in December 1995. So before the Baltimore AAAS meeting I visited one of the two labs that had just demonstrated real quantum logic gate operations, Jeff Kimble's at the California Institute of Technology in Pasadena. Kimble showed me a table where the mysterious mathematics of quantum information became physically real in the form of a maze of mirrors and laser beams. Kimble and colleagues used the elaborate apparatus to make light pulses interact with an

Logic Gates

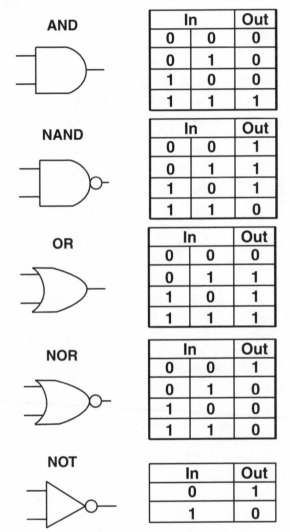

AND		
In		**Out**
0	0	0
0	1	0
1	0	0
1	1	1

NAND		
In		**Out**
0	0	1
0	1	1
1	0	1
1	1	0

OR		
In		**Out**
0	0	0
0	1	1
1	0	1
1	1	1

NOR		
In		**Out**
0	0	1
0	1	0
1	0	0
1	1	0

NOT	
In	**Out**
0	1
1	0

atom of cesium. Laser beams would shoot two photons into a small mirrored cavity, with the photons spinning either to the left or the right (corresponding to the 1 or 0 of information language).

The photons (Kimble called them "flying qubits") bounced back and forth between the two curved mirrors encasing the cavity. The experimenters also shot a beam of cesium atoms through the cavity, so that at any one time one atom interacted with the two photons. After that interaction, two photons emerge from the cavity and are measured to see if their spins have been changed. For the change to occur, both incoming photons must spin in such a way that they interact strongly with the atom. In other words, two photons with different spins will not both interact strongly with the atom, so no change will occur if the two input photons are different. Thus the change in output depends on the input, just what's needed to enable the flying qubit system to do the job of a logic gate.

Meanwhile, researchers at the National Institute of Standards and Technology (NIST) in Boulder, Colorado, had also demonstrated a quantum logic gate, using a rather different approach. They captured a single atom of the metal beryllium in a small electromagnetic trap and then tweaked the atom with laser pulses to conduct quantum logic operations similar to a NOT gate.

Neither the Caltech nor the NIST teams were ready to factor large numbers. To break a code, you'd need to be able to factor a number nearly 200 digits long. Anybody working on quantum computers would be ecstatic to be able to factor a two-digit number like 15. And even that would take more than a single logic gate. To do real computing, you need to put gates together in circuits.

Factoring a big number might take an apparatus trapping a couple thousand atoms, Dave Wineland of the NIST team told me. And it would require lasers that could maintain precise power levels for more than a million pulses. "Lasers we have now wouldn't even approach that," he said.

So the prospects of successful quantum computing remained questionable. After two more years, though, optimism had grown. In early 1998, NASA sponsored a conference in Palm Springs, California, to explore the latest advances in quantum computation. Some of the researchers meeting there expressed growing confidence that full-scale quantum computing could someday be realized. "I don't see any fundamental physical reason why you can't build a quantum computer," said James Franson of Johns Hopkins.

One of the exciting developments by then was the report of a new way of quantum computing, borrowing the principle of nuclear magnetic resonance, or NMR, used by magnetic resonance imaging

(MRI) machines. In this approach quantum information is contained in the spins of atomic nuclei. Those spins are sensitive to magnetic fields. By changing the magnetic fields the quantum information on the nuclei can be manipulated the way laser pulses manipulate the information in trapped atoms.

At the Palm Springs meeting, IBM's Isaac Chuang reported that the MRI version of quantum computing had actually been used to perform simple quantum calculations. A 2-qubit MRI computer solved, for example, a simple version of a database search problem devised by Lov Grover of Bell Labs.

Grover's method showed how, in principle, a quantum computer could search a large database for a single item much faster than an ordinary computer could. Say you wanted to find one name from a list of a million. An ordinary search would take, on average, about half a million tries. (In other words, a computer programmed to look for one name in a million would need an average of half a million steps each time to find the name.) Grover showed that a quantum search could be more efficient. Instead of needing an average number of steps half the size of the list, a quantum search would need only as many steps as (roughly) the square root of the size of the list. Instead of the half a million steps to search for one in a million names in a classical search, a quantum search would need only about 1,000 steps.

Chuang and his colleague Neil Gershenfeld prepared a 2-qubit NMR computer to search a "database" containing four items. A classical search would, on average, need between two and three tries to find the tagged item. The quantum computer could do it in one try. The main issue, of course, was whether the NMR method could be scaled up to commercial size. Scaling up appeared to pose some tough problems. "Whether they'll get to the extent where they're interesting to Intel . . . I don't know yet," Chuang told me.[8] Other experts expressed serious doubts that the MRI approach could succeed on large scales. But shortly after the Palm Springs meeting Vazirani and Leonard Schulman proposed a new mathematical procedure that perhaps would get around some of the scaling-up problems. It remains to be seen. But the situation reminds me of an exchange between Schumacher and the cosmologist Andy Albrecht at lunch during the 1994 Santa Fe workshop. "Experimentalists are able to do some amazing things," Schumacher pointed out. Albrecht nodded. "Basically," he said, "you can't even make a living as an experimentalist unless you routinely do things that a bunch of theorists sitting around at lunch would be completely convinced are impossible."

Even if the NMR approach flops, though, plenty of other possibilities for quantum computing have been popping up. Kimble at Caltech and Wineland and colleagues at NIST continue to tweak their logic-gate schemes in hopes of putting enough gates together to calculate something useful. And Franson has been pursuing a novel approach that might even be able to put quantum computing power into small packages (no doubt this is the approach that Gunter Janek used in *Sneakers*).

In Palm Springs, Franson described experiments on devices about the size of a transistor. Two optical fibers carried input signals into a cell containing a vapor of sodium atoms. Photons sent through the fibers interacted with two of the sodium atoms in the cell, and two fibers exiting the cell transmitted the resulting output. In theory, the photons should be affected in a way that can perform quantum logic, and Franson's preliminary experiments suggested that the effect on the photons predicted by theory is real.

"The effect is definitely there," Franson said. He hopes eventually to be able to replace the sodium vapor with a solid substance like sapphire, improving the chances of successful mass production. "If you can build one, you can build a lot," he said. "It might be possible to mass-produce these at a reasonable cost."

Fragility

Despite growing optimism that one of the many schemes would succeed, none of the new approaches really avoids the problem of quantum information's extreme fragility. The multiple possible positions of an electron or spins of an atomic nucleus can be destroyed by the slightest disturbance. A stray atom or photon flying by can ruin a quantum calculation. And there's always the need for correcting errors—throwing away information—without losing anything.

But many attending the Palm Springs meeting were more optimistic about such problems than Landauer. John Preskill, a Caltech physicist, described a "fault tolerant" approach to quantum computing that would enlist other aspects of quantum weirdness to avoid the problem of outside interference. If some quantum particles share their information with other distant particles, their information will not be so easily destroyed by a stray collision. Of course, Preskill's approach has not yet reached the laboratory demonstration stage. "There were a lot of things that I was implicitly assuming I could do that I don't know how to do," he admitted.[9]

There has also been much additional progress on the error correcting problem. If a quantum computer is designed to transmit qubits redundantly—say, using five qubits to send a single qubit of information—the error can be corrected in the event that one of the qubits is accidentally destroyed. Analyses discussed in Palm Springs show that if quantum hardware can compute with less than one error per 10,000 operations, error-correcting strategies will enable the computer to compute reliably.

Landauer, though, maintained that skepticism remains advisable. "They've won some battles," he said. "I'm not sure they've won the war."

In any case, unless Robert Redford makes a sequel to *Sneakers*, it will be a while before anyone can predict just how important quantum computers will ever be. "Nobody can see far enough to know whether this could change society in the same way [as classical computers]," Kimble told me when I visited his lab. "But the horizons are unlimited because it's so foggy you can't see them."[10]

Chapter 5

The Computational Cell

Biology and computer science—life and computation—are
related. I am confident that at their interface great discoveries await those who seek them.

—LEONARD ADLEMAN,
"Computing with DNA"

When Leonard Adleman had his Eureka moment, it wasn't in a bath-
tub, but in bed.

Adleman, a computer scientist and mathematician, started out
in science in the 1960s as a premed student at Berkeley. He encountered biology back then as "strange things under microscopes," inducing him to abandon medicine for math. It turned out to be a good
choice; Adleman has had a distinguished career during which he
helped invent the standard system for protecting secret codes.[1]

By the 1990s, Adleman's interests had turned to the mathematics of AIDS, so he decided to learn what he could about molecular
biology. But by then, biology was no longer just about strange things
under microscopes. Biology had become the science of how cells use
the information contained in genes.

So when he ventured into a virology lab in 1993, Adleman found
dramatic differences from his student days, he told me when I visited
him in 1998 at the University of Southern California. "I was struck

by the change," he said. "It was just amazing, the difference in what biology had become. . . . It was very clearly now mathematical."[2]

Adleman embraced molecular biology enthusiastically. He began to learn laboratory techniques and embarked on reading the classic textbook *The Molecular Biology of the Gene*, whose authors included James Watson, co-discoverer of the double helix structure of DNA. One night in bed, Adleman was reading in that book about how the two spiral strands of DNA separate to replicate. The master molecule guiding that process is an enzyme that "reads" the sequence of chemical "letters" that encode genetic information along a DNA strand. There are four letters in the DNA code, corresponding to four kinds of "bases"—molecular fragments that connect the two DNA strands. The bases are known by their initial letters: A, T, G, and C, which stand for adenine, thymine, guanine, and cytosine. DNA's two strands stick together because of the way these bases link to each other. A loves to hold hands with T, and G with C. If you find an A somewhere on a DNA strand, you'll know that the other strand has a T at that point, and vice versa.

When it's time for DNA to divide and reproduce, the two strands split and the master enzyme comes along to build each strand a new partner. The enzyme slides along one of the strands, reading the letters one by one. If the enzyme "reads" a C on the original strand, it "writes" a G on the new one under construction. If the enzyme reads an A, it adds a T to the new partner strand.

To the computer scientist in Adleman, this all sounded vaguely familiar. A device moved along a strand of symbols, reading and then writing a new symbol depending on what was just read. Suddenly he realized what it reminded him of. DNA replication was biology's version of a Turing machine.

Adleman sat up in bed. "Jeez," he said to his wife, "these things could compute."

He didn't then run out into the streets in his pajamas shouting Eureka, à la Archimedes. But neither did he sleep that night, staying up trying to figure out ways to make a DNA computer. By 1994 he had succeeded in demonstrating DNA computing in the laboratory. His paper describing that success, published in *Science*, opened a new era of computing.

Adleman's discovery was not exactly the invention of biological computing, however. That honor belongs to the cell itself. It turns out that cells don't need humans to perform computations—cells are full of computational tricks of their own. All of cellular life is involved in transforming inputs into outputs, the way computers do. Cells are not just little bags of alphabet soup, full of things like ATP and NADH, but are tiny chemical calculators. Compared to even the best of human computers, the living cell is an information processor extraordinaire.

"This is a programming system that's been around three billion years," Adleman said. "I'll bet it has a lot to tell us about how to program. . . . It's a dazzling display of information processing ability."[3]

Cells are much more than just computers, of course. They play many roles in life. Cells are building blocks—life is made of cells just as galaxies are made of stars and matter is made of atoms. Cells are factories, making the proteins needed for all of life's purposes, just as real factories make bricks for buildings, vehicles for transportation, and chemicals for doing all kinds of things. The difference is that factories make their products from raw materials and energy. Cells make their protein versions of bricks and buildings by manipulating the bits and bytes stored in the cell's nucleus, in the form of DNA. The DNA in a cell contains enough information not only to make a human body, but to operate one for a lifetime. A gram of dried-out DNA—about the size of two pencil erasers—stores as much information as maybe a trillion CD-ROM disks, Adleman points out.[4] So long before Adleman realized that nature had beaten Alan Turing to the idea of a Turing machine, biologists knew that DNA was the master information storage molecule.

DNA as Hard Drive

Adleman might someday be known as the inventor of the world's most powerful computer. On the other hand, when people of the distant future argue about who invented the DNA computer, some might vote for Watson and Crick. Almost from the moment they figured out DNA's design in 1953, it was clear that it stored the information necessary for life to function, reproduce, and evolve.

DNA's bases are like a series of letters that make three-letter words, and the words are the cell's instructions for making proteins. Each word stands for an amino acid, and proteins are merely long molecules made from hooking amino acids together like links in a chain. The two winding strands of DNA (made of sugar and phosphate) twist around each other a little like the banisters of a spiral staircase, tightly linked because the "letters" on one chain fit like puzzle pieces with the letters on the other chain. (Crick points out that these banisters are technically a helix, which is not exactly the same as a spiral.) A can link with T to make a stairstep connecting the two banisters, or C can link with G, but any other combination won't work—the step would be the wrong size to make a stairstep.

As soon as they figured this out, Watson and Crick realized that the secret of transmitting genetic information had been exposed.[5] "It has not escaped our notice," they wrote in their original paper, "that the specific pairing we have postulated immediately suggests a possible copying mechanism for the genetic material."[6]

The secret was that one of the DNA strands contains the information needed to make the other. Because the bases making steps always combine in the same way, the sequence of letters on one of DNA's two strands determines the "complementary" sequence on the other strand. One strand can serve as a template for making a new partner strand, so DNA can copy itself.

Cells need to copy DNA's genetic information for two reasons: one is to make proteins, the other is to pass the information to new generations so that they, too, can make the same proteins. In this case the new generation can be either a new cell or an entirely new organism.

Making a new cell is the simpler of the two tasks (since sex is not involved). New cells come from old cells; a "parent" cell divides to create two "daughters." When the parent cell divides, its DNA strands must unwind to reproduce—the process Adleman was reading about in bed that night. When the two strands split, each receives a new partner strand, forming two double helices. The cell can then divide in two, each new cell keeping one of the new double helices. That way both daughter cells inherit the same information stored in the DNA of the parent cell.

Getting a new cell is therefore rather easy. Getting a new organism is more complicated, requiring the formation of sex cells by meiosis. After a cell divides by meiosis, each new cell contains only half the normal supply of genes. Meiosis is followed by fertilization, the merging of male and female sex cells to restore a full supply of genetic material. In this process of meiosis and fertilization, DNA from two parents is cut up and recombined, giving the offspring's cells a set of DNA information that contains many similarities while still differing from both parents. In other words, having kids is just complicated information processing.

DNA's other important job—as a blueprint for making proteins—is a little more involved. The blueprint for any given protein is called a gene, and DNA is the stuff that genes are made of. But saying precisely what a gene is isn't easy. It turns out to be much more complicated than the way it's usually presented, as a "segment of DNA." Still, that's the basic idea.

In any case, a gene holds the instructions for producing a protein (or in some cases, other sorts of useful molecules). The instructions are stored in those strings of bases along one of the DNA molecule's strands. The order of those bases in any given gene encodes the information that the cell needs to produce a given protein. (Technically it would be better to say "gene product" since not all genes code for proteins, but let's not worry about that now.)

This basic job description for genes had been worked out by the early 1960s and summarized in what Crick called the Central Dogma—roughly, DNA makes RNA makes proteins. In computing terms, DNA plays the role of a hard drive, permanently storing information. When a given stretch of bases along a DNA chain is "turned on" (itself something of a computational process), another molecule comes along and copies the information off the DNA hard drive onto another molecule, messenger RNA. The messenger RNA plays the part of a floppy disk to carry information outside the nucleus to tiny factories called ribosomes. Ribosomes read the information off the floppy RNA disks to make proteins by stringing amino acids together. There are twenty different amino acids to choose from. The messenger RNA tells the ribosomes which amino acids to connect, and in what order.[7]

The code used by DNA and RNA to transmit this information was not known to humans until the mid-1960s. But it turned out to be fairly simple. As Crick told me when I visited him at the Salk Institute, it *had* to be fairly simple, because DNA's origin was so close to the origin of life itself. When DNA first came on the scene, life hadn't yet had time to get complicated. So once DNA's structure was discovered, progress in understanding its code was really relatively rapid. "What really made the thing was the simplicity of the double helix," Crick said. "It wrote the whole research program."[8]

Information and Evolution

In a way, thanks to DNA and the genetic code, the life sciences adopted the superparadigm of information even before the computer introduced it to the physical sciences. Information and life were a natural fit. After all, science's understanding of life is based on Darwinian evolution by natural selection, and selection is, in essence, information processing. It's all about input and output. In evolution, the environment processes the information presented to it in the form of organisms and produces output—some dead organisms, some live organisms. The information in the live ones survives to make more organisms. DNA, therefore, does not just store information about an individual. DNA is really a record of the selection that has gone on in evolution. Virtually all forms of life, including humans, are descended from their ancestors by the transmission of DNA. So today's DNA is the end of an unbroken thread of information, stretching back to life's beginnings. In other words, DNA is the Herodotus of molecules. For chronicling the history of life, it beats stylus and papyrus, quill and parchment, typewriters and paper, or keyboards and floppy disks.

DNA studies of humanity's remote past have generated considerable controversy. Traditional anthropologists haven't always welcomed unfamiliar molecular evidence. Nevertheless such studies make a compelling case that modern humans originated relatively recently (200,000 years ago or so) in southern Africa, for example. Other studies have traced the migration patterns of early Americans. Reading the history of humanity's ancestors in DNA is not a simple matter. The most common form of DNA, found in the cell's nucleus, doesn't keep the neatest records. Every time parents have a child, the

DNA gets scrambled up in the reproductive process, making genetic history difficult to interpret. Fortunately, the cell has an auxiliary DNA storage system outside the nucleus, in the little structures called mitochondria. Mitochondria contain a small amount of DNA that holds a purer record of the genetic past because it is not mixed up in reproduction—your mitochondrial DNA is almost always inherited entirely from your mother (there are some exceptions). Researchers have used the mitochondrial approach to DNA history to trace the course of human evolution and track patterns in the origins of different languages. More recently, methods have been devised to do similar studies using DNA from the nucleus, based on changes in small snippets of DNA that have not been scrambled in sexual reproduction.

It is, I think, one of the most unexpected developments of the century that molecules within the cells of living humans contain such fruitful information about the history of the human species. For the anthropologists who appreciate it, it's one of the great lagniappes of the information viewpoint. DNA's information storage function alone is reason enough to regard life as in essence an information-processing process. But there's more. Not only can DNA be taught to compute, it computes on its own.

Life as Computation

To explore the issue of cellular computation, I went to Princeton to talk with Laura Landweber, an assistant professor of ecology and evolutionary biology. I found the papers she writes on the subject to be fascinating and insightful, a great mix of creative thought plus mathematical rigor. She exudes enthusiasm for her insights, hopping from point to point with barely a comma between thoughts.

"Biology is itself a computation," she told me. Or in other words, evolution is algorithmic. You can "think about evolution as having years to compute solutions to all sorts of problems—not just life, the universe, and everything,[9] but what is the best life history strategy for a species or what is the best approach for reproductive success of the species, things like that." Evolution, by the standard method of Darwinian natural selection, produces algorithms. "It's a way of letting the algorithms fight among themselves and deem the one which produces the solutions most efficiently or fastest to ultimately win."[10]

Only in the 1990s has this point of view begun to influence biology significantly. "Not many evolutionists have thought about how evolution mirrors the computational processes," Landweber said. But biology itself has known about this for eons. Landweber herself studies squiggly single-celled creatures called ciliates (named for their wispy covering of cilia). It seems that there are at least a couple of ciliate species that beat Adleman to DNA computing by a few million years. Their names are *Oxytricha trifallax* and *Oxytricha nova*.

"These organisms are minute computers," says Landweber. With her collaborator Lila Kari of the University of Western Ontario, Landweber has shown that some of these ciliates perform computational feats equivalent to what Turing machines can do. "They essentially encode read-write rules in their own DNA," Landweber explained.

Understanding the ciliates is instructive, she believes, for such real-life problems as sequencing the human genome, the whole catalog of genetic coding contained in human DNA. If you took the DNA out of a cell and counted all the pairs of bases, you'd find a total of 3 billion of them. But only about 5 percent of that DNA actually contains code for making proteins. The rest is called junk DNA. It's not really all junk; some of it contains instructions for turning genes on and off, for example. But there's still a lot left over that biologists have no idea what it's there for.

"The key point there is that only 5 percent of our known DNA sequences in our genome encode proteins and the other 95 percent is to a large part still mysterious," Landweber said. In the ciliates she studies, the situation is even more extreme—98 percent of the DNA is noncoding in some cases—a further impetus for exuberance.

"They throw out that 98 percent of the DNA," she says, "which is so-called junk, which is beautiful because it's a wonderful solution to this problem of junk DNA, and when you look at the perspective of genome sequencing projects and you realize that 95 percent of the taxpayers' dollars that are paying for genome sequencing projects are sequencing so-called junk DNA, it's great that the ciliates have actually found a way to eliminate their junk DNA, which occupies even more of their genomes."[11]

And how do the ciliates do it? By computing, of course. These ciliates possess not one nucleus, but two. One contains all the DNA, junk and all, in a kind of ticker-tape format, the DNA strung along long chromosomes. From all that DNA the ciliate extracts just the 2 percent of DNA required to make the genes that the organism needs to make proteins, depositing the needed DNA in a second nucleus.

It's a true Turing machine operation—the junk contains the rules for disassembly and extraction. "The key feature," Landweber observed, "is that there's a set of instructions which, like a treasure map, are all encoded in this information which at the end can simply be discarded. But you need it to begin with, otherwise you wouldn't know where to begin."[12]

With every cycle of sexual reproduction the ciliates have to start from scratch, she noted. They begin with a gene that's broken into 52 pieces. And the pieces aren't even all in the same place to begin with. There seems to be a second donor of missing pieces, possibly from a different chromosome. In other words, maybe 40 pieces of a gene are on one chromosome, about 10 on another, and somehow the ciliate has to weave the pieces together to make the proper product. "The final product that's presented to the RNA machinery and the protein translation machinery looks like any other gene," Landweber says.

It's analogous, she says, to the way computer files are stored on a hard disk. Frequently any one file is "fragmented," pieces of it placed on different parts of the disk where space is available. A disk defragmenting program consolidates space by finding all the pieces of various files and grouping them together (to "optimize" the disk's use of space). "So in effect what the ciliates are doing are running one of these optimization algorithms," Landweber said.

It's also a lot like the way e-mail messages travel over the Internet, snipped into packets that travel by various routes and then reassembled by a computer at the receiving end into a coherent message. It's a computational process to put those bits of information back together, just as it's a computational process for the ciliates to take various chunks of genes and merge them into one that makes sense for the cell.

So certain primitive cells can perform some pretty sophisticated computing. But nobody really noticed until computers infiltrated biology laboratories and the thought processes of biologists. And nowadays computers are more important to biologists than microscopes. So it shouldn't be surprising that it has become easier and easier to find examples of computing in biology. In fact, the computational skills of the ciliates are far from unusual curiosities. There's all kinds of computation going on in cells—or at least much of what cells do can be construed in computational terms. Biologists have begun to think not only of DNA as hard drives, but of proteins as Pentium processors and logic gates. If DNA is the cell's computer hard drive, proteins are the RAM, the scene of most of the computing action.

Pentium Proteins

Proteins have many jobs. Some proteins serve as the cell's building materials, while others play the part of the tools that put those materials together. But in the postindustrial cell, proteins have had to learn new vocational skills. (Or more accurately, people have had to learn about proteins' other skills; the proteins knew how to do this stuff all along.)

Fortunately, proteins are very versatile molecules. Their versatility stems from the fact that they come in so many different shapes. When ribosomes make proteins, the task is simply stringing one amino acid after another. Ultimately, when the chain is finished, it might contain a few dozen to several hundred amino acids or more. Finished chains curl up into various kinds of elaborate structures; a protein's precise shape determines what kind of chemistry it can carry out. Enzymes, for example, are proteins shaped in such a way that they are able to accelerate important metabolic reactions (or perform other cellular tasks). If the enzyme is misshapen, efforts at acceleration will fail (and most biological reactions go so slowly without such acceleration that they for all practical purposes don't work at all). If a working enzyme switches to an alternate shape, the chemical reaction essentially stops.

That sounds bad, but it also presents a computing opportunity.

By switching between two possible shapes, proteins can mimic the 0s and 1s of computer language: to represent a 1, the enzyme is active, and the reaction goes; for 0, the enzyme switches to the inactive shape, and the reaction stops. Enzymes thus provide a cell with the basic computing language to process information.

Of course, it's not as if the 1s and 0s can be etched into silicon circuit chips disguised as mitochondria. A cell's interior contains thousands of different proteins. It's not so obvious that such a soup of molecules can actually perform the kinds of digital computations that Pentium processors do. But in fact, humans have been able to show in the laboratory that proteins can compute. It just takes some clever chemistry.

You can think of the cell's chemistry as representing a computer's current memory state, which is why proteins are like RAM. Protein activity changes constantly with new inputs of outside information into the cell. Just as the precise contents of a computer's RAM change constantly as a computer computes, the precise state of a cell's protein chemistry changes to reflect the cell's response to new information. "The imprint of the environment on the concentration and activity of many thousands of proteins in a cell is in effect a memory trace, like a random access memory," writes biologist Dennis Bray of Cambridge University.[13]

A cell's computational skills allow simple life forms to respond to their environment successfully. Bacteria have no brains, after all. Yet they somehow figure out how to swim toward food and away from poison. This behavior is possible because circuits of proteins control bacterial swimming. The bacteria don't have to think about it any more than you do when multiplying numbers on an electronic calculator. A microprocessor does the thinking for you, just as a circuit of protein reactions does the thinking for bacteria. In advanced organisms, with brains, networks of nerve cells process electrical signals to guide behavior. But even the electrical signaling of nerve cells is ultimately a result of proteins changing shape to let charged particles in and out of the cell.

This view of the cell is still not exactly mainstream biology, but it has its adherents. The basic ideas were well expressed by Bray in a 1995 review article in *Nature*. "It must be admitted," he wrote, "that the entire short-term behavior of any organism depends on circuits of proteins which receive signals, transduce them and relay them to neighboring cells."[14]

Other scientists have pursued the notion of cellular chemical computing in great detail. I learned a lot about how cells could compute from the papers of John Ross, a chemist at Stanford. Inside any given cell, he points out, information chemistry can take the form of "cascade" reactions. A chemical cascade is a cyclical series of reactions, in which a molecule produced by one step in the process is needed to drive the next step. Cells conduct a wide variety of cascade reactions responsible for many features of cellular metabolism. "There are many examples of this type of reaction in cells," Ross said.[15]

Enzymes are critical players in cascade reactions—cutting, connecting, and shaping other molecules so they will react. Whether an enzyme is active or inactive determines whether a reaction will produce a lot of a substance or very little. In cascade reactions, the molecule produced by one reaction step can activate or deactivate the enzyme needed to drive the next step. So the various steps of a cascade reaction are all linked.

It's not unlike the way transistors are linked on a computer chip. Transistors are essentially tiny switches. Turned on, they allow a current to pass; turned off, no current flows, just the way a light switch controls whether current flows to a bulb. Transistors are designed to turn on or off based on whether they receive current from their neighboring transistors. In the cell, a step in a reaction cascade will go or not depending on whether the previous step is "on" or "off."

Using computer language to talk about cells is not just a fanciful metaphor. Thinking about a cell's chemistry in a computational way is at the core of explaining how cells decide what to do. Think about it. How do liver cells know when to make sugar, a muscle cell when to contract, a nerve cell when to fire the electrical signal that makes a lightbulb go off in the brain? Ultimately, such cellular behavior depends on which genes in the nucleus have been turned on to make particular proteins. In other words, somehow the cell has to signal the DNA hard drive to activate one of the programs it has stored.

Cellular Logic Gates

In a personal computer, the signal to activate a program comes from the outside world—someone clicking a mouse or typing on a keyboard, for example. In cellular computing, information also enters from the outside world—usually via a "port" made of a protein.

Most cells are studded with protein molecules that act as ports or perhaps sentries, waiting for the signals from outside. These sentry molecules, known as "receptors," poke out through the cell's membrane, sort of like antennas listening for messages. A receptor's job is to tell the cell's insides what information to process from the outside.

Outside messages generally arrive in the form of molecules. The cell's environment is full of such molecules—hormones, neurotransmitters, or other molecular postal workers with information to deliver. Receptor proteins are poised to snatch particular messenger molecules that happen to pass by. Any given receptor is tuned to receive molecules of only a certain shape, much like the way an antenna can be tuned to only one radio frequency. If a molecule of the right shape comes along, it docks on the receptor molecule, fitting, biologists like to say, like a key in a lock. Or a hand in a glove.

A receptor stimulated by the right signal goes into action. As it grips the signaling molecule outside the cell, the receptor changes its shape on the part anchored inside the cell. That change in shape causes proteins on the inside to change shape as well, triggering a chain of chemical reactions that determine how a cell behaves.

Whole circuits of proteins can form to calculate a cell's response to various situations. A signal starting from a receptor in the cell's outer membrane can initiate a cascade of protein reactions that ultimately send molecules into the nucleus. Those molecules may activate genes to produce new proteins in response to that signal from the outside world. As the information from the outside world (input) changes, so does the cell's chemistry and thus its response (output).

Seen in this way, cells obviously share many of the properties of computers. Computers take information in and transform it into output that controls behavior—sometimes just the behavior of your monitor's screen, but often of other devices, anything from a microwave oven to your car's air bag. Cells do the same thing. They take in information from the outside world and compute a chemical response that guides the cell's behavior. In other words, cells guide life not merely by exchange of energy among molecules—that is, simple chemistry—but by the sophisticated processing of information.

G Protein Logic

Okay, so cells don't look much like computers. And you could argue that cells merely respond via chemical reactions to chemical "information," without any real computing going on. But the cell's computational powers actually duplicate some of the principles of logical processing that modern computers use.

Take the heart, for instance. If they think about their heart at all, most people just regard it as an automatic pump. It bears no resemblance to a computer. But heart cells do need to process information about what's going on outside them. The heart must be able to respond to signals suggesting that it beat faster, for example—such as in times of danger or when other stimulation calls for better blood flow. Therefore heart muscle cells must recognize messages from the brain about how hard and fast to contract in order to pump blood. Other signals from the outside, such as the hormones adrenaline and noradrenaline, can signal heart muscle cells to work harder.

Hormones deliver their messages by attaching to specific receptor molecules; the most important in the heart are known as beta-adrenergic receptors. When they flag down passing adrenaline molecules, beta-adrenergic receptors tell the muscles to contract more rapidly and forcefully. (For some people, such as those with high blood pressure, that's not such a good idea, so drugs called beta blockers are used to block the receptors from receiving such signals.)

Naturally, these beta-sentries aren't the whole story of how heartbeat gets enhanced. The receptors are merely the first link in a chemical chain that translates the information about adrenaline's arrival into a proper response by the interior of the cell. When stimulated by adrenaline, the part of the receptor inside the cell signals another molecule, called a G protein. The G protein in turn transfers the message to other cellular molecules. The resulting chemical cascades make the cell contract more.

G proteins play essential roles in most of life's actions—not only in regulating heartbeat, but in seeing and smelling and managing activity in the brain, where the highest levels of G proteins are found. Many of the antennalike receptor molecules protruding outside cells are connected on the inside to G proteins. When such a receptor snatches the appropriate messenger, the G protein goes to work. When they're working properly, G proteins orchestrate biological functions ranging from contracting muscles in humans to mating in yeast. When they go awry, G proteins can be the culprits in such diseases as whooping cough, cholera, and cancer.

Alfred Gilman, now in Dallas at the University of Texas Southwestern Medical Center, discovered G proteins at the University of Virginia in 1977. With his collaborator Elliott Ross, a key player in the discovery, Gilman showed that the G protein is like a switch. The receptor flips it on. It's a very complicated switch, made of three segments. When the switch is off, the segments sit snugly together, with segment A attached to a complicated molecule called guanosine diphosphate, or GDP. (It is the G in GDP, not Gilman, that gives G proteins their name.) When the receptor decides to turn the switch on, the G protein trades the GDP molecule for its close cousin GTP. Now Segment A is free to roam, slithering along the membrane to a second molecule (called an effector). An effector receiving a G protein's message then initiates a whole series of chemical reactions, ultimately leading to the cell's proper response—like making a muscle contract or spitting out sugar. After a few seconds, Segment A turns itself off by transforming GTP back into GDP and returning to its sister subunits to await the next receptor signal.

This picture conveys the basic idea, but is far from the whole story. Recent work has revealed much more sophistication in G proteins' signaling skills. Segment A's two sister subunits are not always passive bystanders, for example, and can play a role in signaling as well. Furthermore, different G proteins can act in concert, activating an effector only if both are on at once. In other situations, one G protein may oppose the efforts of another, so that an effector will work only if one G protein is on and the other stays off. These features are precisely analogous to the logic gates of computers, as Elliott Ross has pointed out.[16]

G proteins typify the value of looking at biology from an information processing point of view. Thinking of the human heart—or heart cells—as computers would not seem to offer obvious medical benefits. You can't replace a diseased heart with an MMX Pentium II. But by understanding cellular information processing, medical researchers can come up with better biological strategies to alleviate medical problems, such as when the beta-adrenergic system breaks down. An example is congestive heart failure. Thanks to the beta-adrenergic receptor information-processing system, the body sends more adrenaline to a failing heart. Or doctors can inject adrenaline or similar drugs into a diseased heart to boost its performance. But after a while, the receptors no longer seem very responsive to the adrenaline message. The number of beta-adrenergic receptors drops, making it harder for the cell to receive the message it needs. And the receptors that remain don't seem to be connected very well with their G proteins. The problem lies not in the lack of adrenaline, but in the flow of information. Animal experiments suggest that gene replacement therapy designed to improve the signaling may someday offer a treatment for congestive heart failure. The search for those strategies is guided by the knowledge that the problem to be solved involves the cell's ability to process information.

There are many other ways that viewing the cell as an information processor could aid medicine. I haven't even mentioned the immune system, which is an exquisite example of information processing. Just as evolution can be viewed as an algorithmic selection process, the body's response to invading organisms is based on identifying information about the invaders and computing appropriate responses.

Another whole area of gene-information interaction involves the way different genes link in networks of activity. No gene is an island; if one is "on," producing a protein, the activity of other genes is affected. And of course some genes produce molecules that directly regulate the activity of other genes. Stuart Kauffman of the Santa Fe Institute has shown how networks of genes can implement Boolean logic functions as some of the genes in a network are turned on and others are off. He draws interesting conclusions about what these computational processes imply for things like the number of different kinds of cell an organism possesses.[17]

To me all this points to the richness of insight that the information-processing view of cellular chemistry provides. Cellular computation is why life's responses to the environment can be so subtle. Mixing chemicals together is generally pretty boring because the same things always happen when the same chemicals are mixed. That's the foundation of the cookbook chemistry of high school and college labs for demonstrating the predictability of the natural world. But life is not so predictable, and is much more interesting, because cells are not merely test tubes, but computers. In a deep biological sense, computing is as much a part of life as eating and breathing.

Out of the Cell, into the Lab

Since it's cellular computing that makes people so smart, it's only fair that people can now make use of the information processing power contained in biological molecules. "It is clear that many of the challenges of manipulating information on a molecular level have been solved by biological systems," writes Peter Kaplan of the University of Pennsylvania with collaborators Guillermo Cecchi and Albert Libchaber of Rockefeller University in New York.[18] It's only natural for envious humans to want to exploit DNA's computing savvy. And for certain kinds of computing problems, it seems, DNA is destined to be preferable to Pentiums.

DNA computing is best at tasks that require trying out a lot of possibilities to find the one that solves your problem. The base-sequence language permits all kinds of chemical editing; molecules can be made to cut DNA chains at specific spots, or to bind two lengths of DNA chain together. So it's possible to make specific chains of DNA to represent quantities or things, and then use editor molecules to conduct computations. If the chemistry is conducted properly, the answer to a problem will appear in the form of a new DNA molecule.

Adleman's original approach, for example, used segments of DNA to represent cities in a variant of the famous traveling salesman problem. The idea is to compute the shortest route among a number of cities, going through each city only once. As the salesman adds cities to his territory, solving that problem gets harder and harder, and soon it's too hard to solve at all. You might be able to find a path through all the cities, but you couldn't be completely sure that there was no shorter path.

In the problem that Adleman tackled, the idea was simply to find a path between seven cities, a tough enough task to see if DNA could really compute. Actually, in his paper, Adleman just used a map with dots labeled by numbers.[19] But it's the same thing as a map with cities. I'll call his "cities" New York, Chicago, Kansas City, San Francisco, Dallas, Atlanta, and Washington. And then he designed specific DNA segments to represent each city. Dallas, for example, was designated by the base sequence TATCGGATCGGTATATCCGA. Then he specified some allowed flight paths. From Dallas, you could fly to Kansas City or New York, but none of the other cities on the map. You could fly to Dallas from Kansas City, New York, or Atlanta. Each allowable flight was also assigned a DNA base sequence (the code for Dallas to Kansas City was GTATATCCGAGCTATTC-GAG).

The flight codes are designed with sequences that are complementary parts of the city codes (their base pairs "match up" and can therefore link the two DNA strands). So when Adleman mixed up these molecules in a test tube, some of them stuck together. A chain of these flight code molecules stuck together would then represent a sequence of flights between cities.

Adleman realized that you could try out a whole bunch of possible flight combinations this way. A mere pinch of DNA for each of the cities and flight codes provided around 100 trillion molecules of each possibility to mingle. In about a second, all the possible combinations of DNA molecules had been created in the test tube.

It then took days for some additional chemistry to weed out the wrong molecules, but Adleman was left with strands of DNA encoding the flight paths between all the cities. It was proof in principle that DNA computing could solve hard problems.

Since then, DNA computing has become a vast research enterprise. Other researchers have explored different DNA computing strategies. One approach uses DNA to make logic gates. Linking different logic gates in the right way makes it possible for a computer to perform all sorts of complicated mathematics.

It's too soon to say whether DNA logic gates will ever steal the computing show from silicon. But University of Rochester researchers Animesh Ray and Mitsunori Ogihara have shown that it's possible to build logic gates with DNA in test tubes. In their scheme, DNA chains of different length represent inputs and outputs. Combining two DNA segments into one longer segment is the equivalent of the AND operation, for example.[20] Chemical methods can be used to search for the long DNA segment to determine whether the AND signal is present.

The Rochester experiments show the sort of versatility that might allow DNA to someday challenge supercomputers for computational supremacy. DNA's power is the power of numbers—it is a lot easier to accumulate trillions of DNA segments than it is to build a supercomputer with a trillion processors. And by using DNA segments just 40 bases long, the Rochester researchers say, they could construct the equivalent of 1 trillion logic gates.

In principle, a full-scale DNA computing center could replace networks of supercomputers for such tasks as predicting the weather or designing airplanes. But not before a lot of sticky problems get solved. DNA chemistry can get complicated. DNA chains can combine improperly to form lengthy chains that confuse the calculations. The chemistry of DNA computing must be designed and controlled very carefully.

But even if DNA computing does not displace the modern supercomputer, it might offer other benefits. Understanding DNA's computational capabilities might reveal unanticipated insights into the nature of life itself. As computer scientist David Gifford commented on Adleman's original paper, "such new computational models for biological systems could have implications for the mechanisms that underlie such important biological systems as evolution and the immune system."[21]

There is, of course, one even more obvious example of computation in biology—the brain. It would be pretty foolish to talk about information processing in biology without discussing the best information processor there is.

Chapter 6

The Computational Brain

Nervous systems . . . are themselves naturally evolved computers—organically constituted, analog in representation,
and parallel in their processing architecture. They represent
features and relations in the world and they enable an animal to adapt to its circumstances. They are a breed of computer whose *modus operandi* still elude us.

—PATRICIA CHURCHLAND AND TERRENCE SEJNOWSKI,
The Computational Brain

And now it's time for the reader participation part of this book.

Hold your hand out in front of your face and spread your fingers
a little. Then keep your head still while waving your hand rapidly
back and forth. Your fingers should become a blur.

Now hold your hand still and shake your head. Your fingers
should seem stationary. (If not, call a neurologist.) For even though
the relative motion of head and hand is the same in both cases, the
brain sees things (in this case, your fingers) differently. It's because
your brain knows how to compute.

When your head moves, your brain knows it, thanks to those
semicircular canals in your ears that help you keep your balance. The
amount of motion measured by your balance system is funneled into

the brain's vision program. As your head goes back and forth in the hand-viewing exercise, the brain's computing system sends a signal to the eyes, telling them to move in the opposite direction—by just the right amount to keep what they're viewing in focus. Without this ability, motion would ruin vision. A wide receiver on the run could never keep his eye on the ball, for example, and football as we know it would be impossible. Walking and chewing gum at the same time would be tough for everybody instead of only klutzes.

Of course, scientists have known for a long time that the brain can compute precise eye-control signals. (It's called the vestibulo-ocular reflex.) But exactly how the brain does it has remained rather mysterious. For that matter, most of the brain's magic is still pretty much a mystery. But there has been a lot of progress in recent years toward demystifying the brain, and a lot of the advances have come from researchers who view the brain as a computer.

Naturally, comparisons have been made between computers and brains ever since there have been computers. Or actually, even before there were computers. Pascal's sister, remember, compared his simple adding machine to the human mind, and the idea of the brain as a calculator was implicit in the comparison. Certainly the first modern digital electronic computer, the ENIAC, evoked the image of a giant brain as soon as its existence was revealed in 1946.

In the 1950s, the person who gave the most serious thought to the brain-computer comparison was John von Neumann, one of the legendary characters of computer lore. Born in Hungary in 1903, von Neumann came to the United States in 1930, already established as one of the world's preeminent mathematicians. From the computer world's point of view, it was one of history's great serendipitous coincidences that von Neumann bumped into Herman Goldstine on a train station platform one day in the summer of 1944. Von Neumann served on the scientific advisory panel for the Ballistics Research Laboratories at the Aberdeen Proving Grounds in Maryland. Goldstine was a participant in the program to build ENIAC at the University of Pennsylvania. Since ENIAC was designed to help calculate ballistic trajectories, Goldstine often found himself in Aberdeen.

Goldstine had never met von Neumann, but recognized him on the Aberdeen train platform while waiting for a train to Philadelphia. "I knew much about him of course and had heard him lecture on several occasions," Goldstine recalled years later. "It was therefore with considerable temerity that I approached this world-famous figure, introduced myself, and started talking."[1]

Von Neumann was friendly, and they chatted in a relaxed way until Goldstine alluded to his work on an electronic computer. "The whole atmosphere of the conversation changed from relaxed good humor to one more like the oral examination for the doctor's degree in mathematics," Goldstine remarked.[2] Shortly thereafter von Neumann visited the ENIAC project in Philadelphia and eventually joined the team. He soon became the world's leading thinker in the theory of computing. It was von Neumann who conceived the idea of storing programs to give computer hardware its instructions for calculating.

Von Neumann himself had a brain like a computer. My favorite von Neumann story involves a problem from computing's early days when a group at the Rand Corporation wanted to know if a computer could be modified to solve a particularly hard problem. They called in von Neumann to explain to him at length why their computer couldn't solve it and to ask his advice on whether they needed a new computer. After they showed von Neumann the problem, his face went blank and he stared at the ceiling a minute or so. Finally he relaxed and said, "Gentlemen, you do not need the computer, I have the answer."[3]

Shortly before he died in 1957, von Neumann prepared a series of lectures (that he unfortunately never delivered) later published as a small book called *The Computer and the Brain*. He had thoroughly studied what was known then about the brain's physiology and of course already knew the intricacies of computers better than any man alive. Computers back then (and even now) cannot do everything a brain can do. But it was clear to von Neumann that the brain was, in its own way, a computer, and he remarked in his lectures that the similarities between computers and brains were even then well known. "There are elements of dissimilarity, too," he noted, "not only in rather obvious respects of size and speed but also in certain much deeper-lying areas."[4]

In particular, von Neumann pointed out that the brain was only partly digital. It was also part analog—that is, that math is carried out not by a direct calculation with numbers (or digits) but by a physical process. Nowadays the difference between digital and analog is rather widely known, but in the 1950s, I suspect, those terms were a little more unfamiliar. Even today most people would probably have a hard time explaining the distinction, although they could give examples, typically citing clocks or watches. Clocks, after all, are computers in a restricted sense—they compute what time it is—so they illustrate the difference between digital and analog computing pretty well. A clock face or watch with hands is analog; if it just shows an electronic display of numbers (that is, digits), it's digital. Clocks with hands are analog computers because digits are not computed directly; the numbers are represented by the physical movement of the hands. The physical process is an analogy to the numerical process.

Another analog example is the slide rule, a set of sliding sticks that can be found today in museums or elderly scientists' attics. Distances marked out physically on a slide rule represent numbers. Calculations are conducted by moving the parts of a slide rule different physical distances, just as an analog clock computes the time by moving its hands a certain amount to reflect a given amount of time. Digital calculations, on the other hand, manipulate numbers directly by devices that perform the logical operations that underlie arithmetic. In von Neumann's terms, analog computing is physical; digital computing is logical. Or as the physicist and science writer Jeremy Bernstein once summed it up, "an analog calculator measures, while a digital calculator counts."[5]

Von Neumann's analysis of the nervous system persuaded him that at least on the surface, brains behaved very much like digital computers. Nerve cells (or neurons) work by firing electrical impulses. It is easy to imagine an impulse as representing a 1, and the lack of an impulse a 0, in the binary language used in computer logic. Furthermore, von Neumann noted, whether a neuron fires an impulse or not depends on the signals it receives from other neurons to which it is connected. Only a certain number or combination of input pulses will induce a given neuron to fire a pulse of its own.

"That is, the neuron is an organ which accepts and emits definite physical entities, the pulses," von Neumann wrote. "Upon receipt of pulses in certain combinations and synchronisms it will be stimulated to emit a pulse of its own, otherwise it will not emit. . . . This is clearly the description of the functioning of an organ in a digital machine."[6]

Von Neumann was fully aware that this description was simplified and that the digital nature of the brain was not really so clearcut. For one thing, he realized that the brain used chemicals to represent information in a physical (or analog) way, and that the brain's operation must involve some switching back and forth between digital and analog components. Nevertheless, it was clear to von Neumann that the brain does in fact compute. Nowadays this is a common notion. "The brain computes!" exclaims the computational neuroscientist Christof Koch. "This is accepted as a truism by the majority of neuroscientists engaged in discovering the principles employed in the design and operation of nervous systems."[7]

For some reason, though, this is not a message that everybody appreciates receiving. It is the flip side of another contentious coin, the question of whether computers can think. A lot of people don't think so—or perhaps more accurately, don't want to believe so. I think some people don't like the idea that the brain is a computer because if a brain can compute (or if a brain works its magic by computing), then it seems somehow harder to deny that a computer can do what the brain does.

To some degree this is a meaningless argument. The fact is that within the last decade or so, computers have become so powerful that comparisons to the brain have become more and more meaningful— and more useful. Computers are now one of the neurosciences' favorite tools for investigating how the brain works. Computer models can help explain not only why your fingers stay still when you shake your head, but also how the brain generates other motions, responds to stimuli, and learns. Even brain diseases like epilepsy may someday be better diagnosed and treated with the help of computer models of mental processes gone awry. In the end, it should be obvious that brains process information, and it makes perfect sense to describe information processing in computational terms.

Representation of Information

Nobody should be surprised that the brain has a lot in common with computers. Brains are made of cells, and we've already seen how cells themselves are full of computational molecules. But brains give life a new information-processing dimension, beyond what goes on inside cells. After all, if cells can compute, then why are plants so dumb? Plants do possess cells that conduct some pretty sophisticated information-processing chemistry. Nevertheless, on the whole, plants aren't so smart—because plants are like the Scarecrow in *The Wizard of Oz*. They have no brain.

Plants have no brain because their ancestors made their own food. Of course, making your own food isn't all that easy—photosynthesis is pretty sophisticated chemistry. But it doesn't require a brain. Eating food is the more complicated task. It requires a sophisticated system to control the muscular motions enabling an organism to see and catch edible life forms. The nervous systems used by primitive life for eating were probably the first step on the evolutionary path to a brain. In other words, we eat, therefore we think.

Computers, on the other hand, don't eat, but they do think—sort of. So the comparison of computer thinking to human thinking is not entirely parallel. Yet computers and brains both process information, and that is the key point. Ira Black, a neuroscientist at the Robert Wood Johnson Medical School in New Jersey, articulated this idea a few years back in a concise, insightful book called *Information and the Brain*. Both computers and the brain process information by manipulating symbols, he pointed out. In computers, the symbols are electronic patterns. In brains, the symbols are molecules.

Black asserts that the brain possesses the defining properties of an information processor—it translates information from the environment (the input from the senses) into molecular symbols that carry messages (output). Molecules produced in response to input can carry chemical messages to the genes inside the brain's nerve cells, telling them what molecules to produce in what amounts. (In other words, activating programs stored on the neuron's DNA hard drive.) Different environmental input induces different chemical reactions that activate different DNA programs.

But there's an important difference between brains and computers, Black points out, involving the distinction between software and hardware. Computer hardware sits around doing nothing unless software is activated. In the brain, there is no real distinction. Molecules are both the symbols representing information in the brain and the tools that perform the brain's tasks. So hardware and software are one and the same. The brain is in no way a hard-wired machine with pre-fabricated circuitry, but a flexible and complicated chemical factory, constantly reorganizing itself in the light of new experiences.

"In contrast to the idea of a fixed, digital, hard-wired switch-board, we're now fully coming to appreciate that there's extraordinary flexibility" in the brain, Black says.[8] That flexibility is reflected in the brain's complex chemistry, which produces molecules that can serve as a memory of an experience. Even a brief experience, lasting seconds to minutes, can induce chemical changes in the brain lasting days to weeks. And the memory molecules may have another job, namely responding to the experience. For example, in parts of the nervous system, stressful signals from the outside world activate an enzyme called tyrosine hydroxylase. It happens to be a key player in the body's fight-or-flight response. So elevated amounts of this molecule perform two jobs: representing information about stress in the environment and playing a physiological role in the response to that stress.

Human behavior, Black asserts, can ultimately be viewed as the output of all this molecular messaging. He sees no problem in principle in describing all brain processes as part of a chemical loop connecting genes and environment. Notions of "mind" and "self" should emerge naturally from understanding this loop, he says, which encompasses all the levels of the nervous system—from chemical reactions among biological molecules in cells to physical activity to abstract thinking. "In principle . . . the apparently vast gulf between molecule and behavior can be bridged," he wrote in his book.[9]

There's a long way to go before construction of that bridge is finished. The environmental impact statement hasn't even been written. Yet scientists have gone a long way toward showing how the brain's biology can be understood in terms of information processing. At the frontier of this approach are the practitioners of a relatively new field of study called computational neuroscience.

Computational Neuroscience

Computational neuroscience has something of a dual personality. On the one hand, computational ideas can help scientists understand how the brain's nerve cells conspire to create thought and behavior. On the other hand, understanding how the brain computes can give computer scientists clues about how to build better computers. After all, for complex tasks (like chewing and walking simultaneously), the brain's programming is apparently more sophisticated than anything Microsoft can manage.

"Things that we thought were very difficult and incredibly intellectual—like playing chess or medical diagnosis—can now be done on your PC with simple programs," says neuroscientist Terry Sejnowski, "whereas something as simple as walking and chewing gum at the same time is something that nobody has written a program for."[10]

Sejnowski is one of the chief intellectual instigators of the computational approach to neuroscience and coauthor of a thick and erudite book on the topic called *The Computational Brain*. He's a deep thinker and fast talker who exudes intensity and intelligence. He has helped put computational neuroscience on the map in California at the Salk Institute for Biological Studies in La Jolla, one of the half dozen or so greatest institutes of intellect in North America.[11]

Sejnowski's field of computational neuroscience is a little hard to define. Basically the field is "computational" in two senses—using computers to make models of neurons, and using those models to study how neurons compute. Neurons aren't the whole story, though. They don't work in a vacuum. To explain how neurons generate behavior, scientists must study how neurons link up to process information in networks. You've probably heard of "neural networks," collections of artificial neurons connected in grids or lattices. (Often there is no real grid, just the simulation of a grid in an ordinary computer.) In any case, neural networks are based loosely on a simplified notion of how the brain works. In a basic neural network any individual neuron is either "on" or "off," corresponding to a real neuron's either firing an electrical impulse or remaining quiet. (An average neuron receives signals from thousands of others; whether that neuron fires a signal of its own depends on the total effect of all the incoming messages, as von Neumann pointed out.)

But the artificial neurons in neural networks are nothing like real neurons. Though they have found some practical uses, neural networks generally have very little to do with the brain. Computational neuroscientists, though, have begun to take neural networks to a more sophisticated level, with more realistic computer simulations of how real neurons work. Such computer models can mimic how a neuron's output depends on input from a multitude of messengers. These simulations help scientists see how information flows through the brain's loops of interconnected neurons to process input from the senses, store memories, and orchestrate thought, muscular motion, and other behavior.

Of course, understanding behavior by studying neurons is a little like learning how to drive a car by studying the chemistry of gasoline combustion. There's a lot in between gasoline burning and driving that you need to know to get a license. Ultimately, understanding the brain will require a whole ladder of understanding, leading from the activity of neurons to the actions of whole organisms.

And for that matter, says Sejnowski, you have to start at an even more elementary level than the neuron. At the bottom of the ladder are individual molecules, such as the messenger molecules that shuttle from neuron to neuron and the proteins embedded in neuron membranes—those antenna or receptor molecules. The G protein logic gates play particularly prominent roles in many kinds of neurons, and the brain's computational power begins with the computational abilities of its individual cells. More advanced computational powers arise from how the brain's cells are connected. So the next step up the ladder is the synapse, the transfer point where molecules jump from one neuron to another. Then come small networks of connected neurons, and then the more elaborate subsystems that make up the whole central nervous system. Understanding behavior, says Sejnowski, will require knowledge of all of those levels. And he thinks that knowledge can come from constructing computational models to simulate each step in the ladder.

So far, much of the modeling by computational neuroscientists has focused on the single neuron, trying to capture its internal complexity and its electrical properties. Older approaches viewing a neuron as a simple on-or-off switch, firing an electrical impulse when stimulated but otherwise dormant, don't really resemble the real thing. Most neurons, for instance, fire at a "resting rate" no matter what is going on. When stimulated, the neuron will then fire at a much more rapid rate. In many cases, just as important as the firing rate is the precise time of firing, particularly whether the firing of one neuron is going on at the same instant as the firing of other neurons. "Time is really a neglected dimension," says Sejnowski. "It's the next frontier."[12]

Someday, neuroscientists hope, devices will be available to study the brain with exquisite resolution, both in space and time, so that the exact timing patterns of individual nerve cells at precise locations can be analyzed in real life. In the meantime, computer simulations provide the best available way to study such issues. And the computer simulation approach offers many advantages over traditional neuroscience methods. For one thing, many experiments are a lot easier to perform on a computer screen than with tiny neurons in a laboratory dish—or in a brain. A computer model can be manipulated with a lot less mess. And the model can provide a more complete picture of what's going on in cases where direct experimental evidence is muddled or sketchy.

In testing the effects of drugs on the brain, for example, the traditional approach is to administer the drug and see what happens. But that method offers little control over where the drug goes. It's hard to tell the primary effect from indirect side effects. Or as Sejnowski puts it, "It's like taking a sledgehammer and hitting a computer. The screen may go blank, but you don't really know exactly why."[13]

Much of modern medicine has been developed by this hit-or-miss approach. But making further progress on some diseases may be easier with a computer-assisted strategy. Neurologist William Lytton of the University of Wisconsin–Madison, for example, has worked on computer models to better understand the effects of epilepsy drugs on the brain. Feeding knowledge from real experiments into the computer allows the model to isolate the roles of different kinds of neurons in epilepsy and help predict which drugs offer the most potential benefit.

More recently, scientists have succeeded in growing real nerve cells (from rat brains) on silicon chips. That allows precise monitoring of how the real cells send signals, and permits the scientists to inject signals of their own to influence the nerve cell network. This work is the first step toward the science-fiction future with computer chips implanted in living brains—possibly to treat neurological disorders, or to permit mental capabilities beyond the imaginations of brains without silicon in them.

Sejnowski and other computational neuroscience pioneers have faith in their field because it helps them find out things they didn't already know, or couldn't have figured out otherwise. In other words, computational neuroscience offers the element of surprise—as has happened in studies of how the eyes compensate for the motion of your head in the finger viewing experiment. Using computer models to explore that phenomenon has led to a totally unexpected explanation of how the brain computes the instructions to the eye muscles, for example.

That eye-motion skill, or vestibulo-ocular reflex, has been studied extensively by Sejnowski and Stephen Lisberger of the University of California, San Francisco. They tried to figure out how the reflex works by changing the nature of the task. Monkeys, like people, can keep an image in focus while moving their heads. But Sejnowski and Lisberger changed the game by placing goggles on the monkeys to reduce or enlarge visual images. When the image size changes, the eyes must move by a different amount to compensate for head motion and keep the image in focus. Typically it takes a few days for the brain to adjust to the goggles and keep objects clear while the head is moving. Apparently the brain learns new computational rules during that time.

Traditionally, neuroscientists have believed such learning must involve changes in the synapses, where neurons communicate. "Everybody who has thought about this problem has always assumed . . . that the synapses are changing their strength," said Sejnowski. "That's almost a dogma."[14] And in fact, Sejnowski and Lisberger began their modeling of that process using the synapse assumption. But then the model steered them in another direction. Playing around with the computer model, they found that the eye-motion reflex could be modified by changes in a nerve cell's gateways for the electrically charged particles known as ions. These gateways, or ion channels, control the flow of electrical current in neurons. The speed with which the channels open and close can control the eye's response to head motion. The important change required to adjust to the goggles seems to come not from changes in synapses, but from changes in ion channels.

The implications go beyond hand waving or catching footballs. In a paper on their findings published in *Nature*, Sejnowski and Lisberger noted that the eye reflex adaptation may be just one example of how subtle changes in individual cells can affect behavioral skills. Studying the brain circuits and feedback mechanisms involved in this reflex, they suggested, could lead to better understanding of many other processes linking neuron activity to behavior.

Does the Brain Really Compute?

Despite all this progress in computational neuroscience, the argument about whether the brain is a computer has not gone away. There is always somebody around who when talking about the brain sounds like the robot from the TV series *Lost in Space*, saying "It does not compute."

Perhaps the most vigorous and elaborate denunciation of the brain as computer has come from Roger Penrose, the British mathematician who articulated his case in two popular books—*The Emperor's New Mind* and *Shadows of the Mind*.

Penrose insists that the brain is not a computer because it can do things that a Turing machine cannot. For instance, it is well known that some mathematical statements cannot be proven true within a given mathematical system of rules and axioms. A Turing machine would run forever without being able to answer whether such a statement is true. But a human can figure out the truth of some such statements anyway.

For example, take a mathematical statement that can be translated into a sentence—call it sentence X—that means "this sentence can't be proved." Then assert a second statement, call it statement Y, that "there is no proof (within this axiom system) of sentence X." In other words, "There is no proof that X can't be proved."

A Turing machine would spin its gears forever trying to prove that there is no proof of an unprovable sentence. But a human brain can see that statement Y must be true. Look at statement Y again: "There is no proof that X can't be proved."

Suppose sentence X could, in fact, be proved. Then statement Y is still correct, because if X *can* be proved, it is obvious that "there would be no proof" that it *can't* be proved. On the other hand, if X *can't* be proved, the statement is again correct, because to say "There is a proof" that X *can't* be proved is self-contradictory. Therefore, statement Y is true, even though a Turing machine could not prove it true.[15]

So people can somehow see that certain problems can't be solved, while a Turing machine could run on forever without being able to decide that those particular problems were not solvable. Penrose takes a great deal of comfort from this supposed superiority of humans to machines, and he concludes that consciousness is not computational. But to make his case, Penrose combines a lot of ideas from different contexts, and it's not clear to me that all his cards come from the same deck. I've read Penrose's books carefully, and I talked to him at length about some aspects of his argument before his first book had been published. I thought his ideas were intriguing but somehow not compelling.

Since consciousness gets its own chapter (coming up next), Penrose will appear again later. For now let's just say I have a hard time buying the argument that the brain is not a computer. The simple hand-waving test reveals—or so it seems to me—that brains clearly do, in some sense, compute. Computers represent information, that is, store it in memory, and manipulate information, transforming the "input" signals into "output"—new memories or new signals. It seems natural enough to describe what the brain does in similar terms.

Nevertheless, Penrose isn't alone—various philosophers also object to calling the brain a computer. Computing may seem similar to thinking, but many philosophers say it's not obvious that people think the same way that computers calculate. To that I have to agree; thought remains largely mysterious. I just don't see why that matters. Airplanes and birds use different technologies, but they both fly. It's perfectly reasonable that both brains and silicon chips could compute.

One of the most famous philosophical attacks on the brain as computer (cited approvingly by Penrose) is the "Chinese room" thought experiment devised by the philosopher John Searle. He poses a clever scenario that boils down to saying people understand what they're doing and computers don't. It's wrong, Searle insists, to say a computer "understands" anything the way a human brain does, because a computer just follows the rules it is programmed with.

Searle illustrates this point with a scenario that has been described many times in discussions of artificial intelligence, the field involved in trying to make "thinking machines." To Searle, though, "thinking machine" is an oxymoron. To make his point, he imagines putting a man who knows no Chinese in an empty room, sealed off from the outside world except for one channel (say a door) to be used by the experimenter. The man inside would be given a bunch of cards containing Chinese symbols and a book of instructions (in English). The instructions (like a computer program) tell the person what symbols to send out in response to Chinese messages slipped under the door.

When the experimenter gives the man a Chinese message, the man consults his list of rules and passes back a card with a symbol as the rules dictated. To the outside observer, the man could appear to be responding appropriately to questions posed in Chinese symbols (if the rules were properly written) even though the man himself would have no idea what he had been communicating. To Searle, this illustrates the rule-based input-output system of a computer, and demonstrates that the computer does not understand what it does.[16]

I once interviewed Searle, and it was rather challenging to follow his arguments. He literally wouldn't sit still to be interviewed—in fact, he wouldn't sit at all, but rather paced around the room in quasi-random fashion, running one long sentence into another—but very articulately. He seemed to be saying that the brain couldn't be a computer (or maybe it was that a computer couldn't be a brain) because a brain is "biological."

Searle's point was that a computer could not reproduce a "mental state" like consciousness because mental states are inseparable from the biological system (that is, a brain) producing them, whereas a computer just runs a program, a set of formal logical instructions that are independent of the machine that runs the program. "I think I've refuted the view that computers can have mental states just by virtue of running a program," Searle told me. "We've got to get out of the idea that mental means nonphysical and physical means non-mental. And the ultimate form of this is the idea that somehow consciousness is not part of the ordinary biological world. I want to say consciousness is as much a part of biology as digestion or photosynthesis or the secretion of bile." He conceded, though, when I asked him whether ultimately understanding the missing biological principles underlying consciousness would make it possible to build a "conscious" machine. "Sure," he said. "If we had a perfect science of the brain, so we really knew how the brain did it, I don't see any reason in principle why you couldn't build an artificial brain that would do the same thing our brain does, namely produce consciousness."[17]

Probably dozens of refutations of his Chinese room argument have been written by computer scientists. My favorite comes from the philosophers Paul and Pat Churchland, philosophers at the University of California, San Diego, who point out that the analogy to the brain could be made much clearer if you put many men (let's say 100 billion) in the room and let them work together to respond to the Chinese symbols. The output of the room will appear to be sensible, signifying some understanding, even though no one man understands Chinese.[18] So how is this different from the brain, where 100 billion neurons cooperate to give sensible sounding output, even though no single neuron has the slightest idea what *any* human language means?

Another way to look at it, it seems to me, is that Searle's Chinese room lacks the kind of sensation and feedback it would take to make sense of the outside world. I think given a wide enough range of input and feedback, the man in the room eventually would learn to understand Chinese, just as a human baby eventually learns whatever language it grows up with. Regardless, it doesn't strike me as a very sound argument that brains can't, or don't, compute. Depending, I guess, on what you mean by compute. In other words, the issue may boil down to mere semantics.

That's the way Sejnowski saw it when I posed Searle's objection to him.

"I think that some of this problem is semantic," Sejnowski said. "If you push him about what he thinks is actually going on in the brain, he says well there's some biochemistry going on. Okay, whether you call that computational or not is just semantic, right? I think it's computational because I'm describing it with equations . . . and it carries information. So why not call that computational?"[19]

Part of the issue, Sejnowski says, is that the word *computer* evokes images in most people of a digital machine with a silicon chip. It's true that the brain doesn't fit that image—brains were around, after all, long before Macintoshes. And before there were digital computers, people computed with different devices like slide rules that used analog computation. It wasn't digital but it was still computing, just analog computing. "And that's what you see in the brain, so why can't you call that computing?" Sejnowski said. "I think it's completely silly to argue about terminology like that."

Basically I think Sejnowski is pretty much right about this. It's the duck argument—if a brain quacks like a computer, it's a computer. Still, I think there is more to the argument than semantics, especially when getting into the issue of just how the brain does all that computing.

Dynamical Debate

I don't want to get into one aspect of this issue—namely, the issue raised by critics of the computer model who simply object to any attempt to describe the mind's mental powers in physical terms. Spiritual approaches to mentality carry the discussion outside the realm of science, and I don't want to go there. Within science itself, though, is another argument about which sort of physical mechanism is at work in the mind: a computational mechanism, or a dynamical mechanism.

The "dynamical hypothesis" suggests that the mind's powers stem not from mimicking microprocessors, but from the natural processes of molecules in motion. From this point of view, thought is just like other aspects of nature. In a dynamical system, the parts move around in different paths governed by the equations describing motion, the way the planets orbit the sun in accordance with Newton's laws. Change, in other words, results from matter in motion.

The solar system is, in fact, the prototypical dynamical system, easily comprehended because of its simplicity. But most dynamical systems are much more complicated—with countless tiny particles, atoms and molecules instead of planets, whirring around under the influences of external forces and their mutual interactions. The typical example of a complex dynamical system is the weather.

Now, it's true that weather phenomena seem to capture the way that some people's brains work. There are such things as brainstorms, for example, where ideas pop out of thin air like lightning in a thunderstorm. And there are people whose minds seem as turbulent as tornadoes. But many scientists insist that thought is not in its essence meteorological. Thinking deals with information, as do computers, and therefore thought involves manipulating symbols that represent information. That makes thought computational, not dynamical. This is the issue, it seems to me, that captures the essence of the argument. Is thought, or more broadly consciousness itself, merely the result of matter in motion, or is it a process of representing and manipulating information?

Put another way, is what the brain does real computation or metaphorical computation? I like to think it is real computation. But it never hurts to get a second opinion. So I asked Rolf Landauer, the prophet of information physics, about the fairness of extending the idea of computation to processes in nature other than electronic computers. In effect, I was trying to negotiate his approval for what I wanted to say anyway.

"People occasionally stretch the definition of a computer to the point of meaninglessness," Landauer said. "To me a computer is something which you load ones and zeros into it, you get going and you get ones and zeros out at the end of it in a very similar format, ready if necessary to put into the next computer program. There's a standardized format for the input and the output. Once you back away from that I think it loses all meaning. There are a number of papers which more or less equate any physical process, any evolution of a system, to a computation. And I think that's giving up on the meaning of computation."[20]

He cited papers, for example, saying that any quantum mechanical process is really a computational process.

"I think that's stretching the interpretation of the word *computation* too far," he said. Whereupon I began my interrogation:

ME: Is that stretching it too far in the sense that it's no longer real computation but is maybe metaphorical computation?

LANDAUER: I don't know how to answer that.

ME: What about something biological like the brain, which takes sensory input and translates it into some motor output . . . ?

LANDAUER: I'm not commissar of what's called computing.

ME: But is that an example of what you have in mind about taking the idea of computation too far?

LANDAUER: No. No, I don't think so.

ME: That's legitimate?

LANDAUER: Well, it takes more thinking; it's borderline.

ME: But that concept is a legitimate concept of computation, it's got the input of data and the output, and that's not as bad —

LANDAUER: That's not as bad as taking some quantum mechanical gadget which is just going and changing its trajectory over time. . . . Now you get to a matter of taste, what you call a computation. It doesn't offend me.

ME (silently, to myself): Whew!

Landauer's less than wildly enthusiastic acquiescence made me want to search more deeply into the rationale for thinking of the brain as a computer. Comparing dynamics to computation can get pretty complicated. It turns out that the question is not all that clear-cut. Ultimately it involves a debate that drags us, kicking and screaming, into the controversial world of chaos and complexity.

Chapter 7

Consciousness and Complexity

Perhaps the most remarkable property of conscious experience is its extraordinary differentiation or complexity. The number of different conscious states that can be accessed over a short time is exceedingly large. . . . The occurrence of a given conscious state implies an extremely rapid selection among a repertoire of possible conscious states. . . . Differentiation among a repertoire of possibilities constitutes information, in the specific sense of reduction of uncertainty. Although this is often taken for granted, the occurrence of one particular conscious state over billions of others therefore constitutes a correspondingly large amount of information.

—GIULIO TONONI AND GERALD EDELMAN,
"Consciousness and Complexity"

Every field of human endeavor has had legendary collaborators.

History had Romulus and Remus, Lewis and Clark, Stanley and Livingstone. Music had Gilbert and Sullivan, Lerner and Loewe, Lennon and McCartney. Baseball had Ruth and Gehrig, Mantle and Maris, and now McGwire and Sosa. Magic has Siegfried (no relation) and Roy. Fiction has Holmes and Watson, Zorro and Bernardo, and the dynamic duo of comic books and film, Batman and Robin.

Science has Watson and Crick.

They're not a duo anymore, of course. It was almost half a century ago that James Watson and Francis Crick figured out how DNA is put together, ultimately leading to the modern understanding of heredity and the genetic code. Afterwards, Watson stuck with molecular biology. Crick had some unfinished business with the brain.

In retrospect it is lucky for science that Crick chose to pursue his interests in the proper order. He was originally trained in physics and spent World War II applying his knowledge to making mines designed to sink German ships. After the war he wanted to pursue a career in research—but not in physics. He realized that the things he liked to gossip about were the latest developments in molecular biology (in those days, biochemistry) and the science of the brain. It seemed logical to him that he should choose to study one of those two fields.

When I visited Crick in 1998, half a century after he'd faced that choice, I asked him if he had decided on molecular biology because it was riper for discovery then than neuroscience was.

"No," he said. "It was because . . . my background was nearer to what we now call molecular biology. . . . I didn't know enough about either subject. I was learning but my knowledge was very fragmentary. But I thought, well, look, I have a training in physics and I know a bit of chemistry; I don't know anything about the brain."

So he decided to take the path to the double helix. But first he had to avoid a detour that would have distorted biological history. A few weeks after his decision, he was offered a job to do vision research.

"It wasn't even a studentship, it was a job," he said. "You know, with a salary. And I had a big struggle and I eventually thought well, you know, you decided that this was the right thing to do and you shouldn't be deflected by an accidental offer of a job."[1]

So Crick stayed to the path and became a molecular biology legend. Many years later, with no more genetic codes to conquer, he turned back to his other favorite gossip subject, the brain. With the secret of heredity revealed, why not solve the mystery of consciousness? And where best to start the study of consciousness? By investigating vision, he decided, the topic of the job he had long ago turned down.

Pursuing the mystery of how the brain orchestrates visual awareness is what Crick has been up to for the last two decades at the Salk Institute in La Jolla. There have been no breakthroughs comparable to the discovery of the double helix. In fact, Crick says, the question of consciousness is still on the frontier of neuroscience, in the sense of frontier as a sparsely populated wilderness.

"There are two points of view among scientists" about consciousness, Crick said. "One is that it isn't a scientific question and is best left to philosophers. . . . And the other one is that even if it's a scientific question it's too early to tackle it now. And there's a fraction of truth in both of those."[2]

Consciousness should be studied scientifically only within a broader program of research that attempts to understand the whole brain, Crick believes.

"It wouldn't be logical at all for everybody in neuroscience . . . to work on consciousness," he said. "I think what one really should say, if you're trying to look forward, is that it's the study of the brain as a whole which will be the thing for the coming century. . . . Much of the activity of the brain, we suspect, is not associated with consciousness. And if you understood a lot of that activity it might help you to discover the mechanisms that are there, some of which would also be used for consciousness."[3]

Crick's consciousness collaborator, Christof Koch of Caltech, is a leading computational neuroscientist. Their efforts to understand visual awareness, as an essential aspect of consciousness, have a distinctly computational flavor. But as Crick emphasizes, not enough is yet known about the brain to say for sure how its ultimate explanation should be framed. It would surely be a mistake to assume that consciousness can be described simply in terms of computation, especially as debates continue about whether what the brain does is computation at all. Understanding consciousness will not be simple. It will require coming to terms with the brain's complexity. And dealing with complexity is something that science is just now learning to do.

Dynamical Dualism

The guiding goal behind the study of complexity, naturally, is to find a way to describe what goes on in complicated systems. The brain is a perfect example. But so are lots of other things, like the weather. Even the solar system, though far simpler than the weather or the brain, has enough objects spinning around to make it complicated. And in fact, it was an investigation of the solar system that led to the modern way of describing complicated things as "dynamical systems."

This idea was developed by Henri Poincaré, the French genius who not only mastered every area of mathematics, but invented a few new ones. Born in 1854 in Nancy, France, Poincare earned his math doctorate in 1879 and two years later took the chair in mathematical physics at the Sorbonne. Not only did he develop new fields of math, but he also found time to investigate many fields of physics, even laying the groundwork for some of Einstein's theory of special relativity. Poincaré was, as E. T. Bell wrote in *Men of Mathematics*, "the last universalist."[4]

When it came to studying the solar system, Poincaré was inspired by a contest sponsored by the king of Sweden. The contest included the question of whether the solar system was stable—that is, would planets continue forever in their orbits, or someday fly apart? Poincaré could not answer the question definitively, but his analysis of the problem was so insightful that he won the prize anyway. Out of that analysis emerged the concept of dynamical systems.

In this case *dynamic* doesn't merely mean active and forceful, like Batman and Robin; it refers to a system described in terms of the motion of its parts, motion governed by the equations of physical law. The basic idea behind dynamical systems is describing how a system changes over the course of time. Typically scientists describe this change by specifying the location of all the particles in a system and how fast they are moving (and in what direction). The equations that describe the motion of the particles can then, in principle at least, be solved to tell where the particles will be in the future, based on where they all are now. (This is how astronomers can predict eclipses far enough in advance to make vacation plans at an optimum-viewing location.)

Poincaré realized that most dynamical systems are chaotic. That means it's impossible to predict what they will do, precisely, very far into the future. Even though it might be possible in principle to predict the future exactly, such a prediction would require absolutely accurate information about the current conditions in a system, an unattainable requirement. Now, if a small error in the initial conditions caused only a similarly small error in the prediction for the future, there would be no serious problem. But in a chaotic dynamical system, a small change in original conditions can produce a large and unpredictable difference at a later time. The standard textbook example is called the butterfly effect—flapping wings in South America can cause a thunderstorm in South Carolina. (I prefer a baseball illustration—a swing and a miss by Sammy Sosa in Chicago can cause a baseball game to be rained out in Cleveland.)

A simpler example (studied to death by chaos scientists), based on a similar principle, is a sandpile. Adding a single grain of sand (surely a small change) might just make the sandpile a tiny bit bigger, or it might trigger a massive avalanche. (Or it might trigger a small avalanche. The distribution of different sizes of avalanches in a sandpile is itself a subject of extensive study.)

Poincaré couldn't make much progress in studying the details of dynamical systems, because the computations were much too complicated for paper and pencil. Real progress came only with the modern computer. Advances in computer power and speed have nowadays made it possible to describe complex and chaotic systems like the weather with some degree of mathematical rigor. In essence, modern computing power has turned the study of dynamical systems into an industry, making "chaos" a serious field of scientific study rather than the description of some scientists' offices. As a result, the dynamical approach to describing nature can now be applied to problems that scientists of the past would have considered untouchable—such as the nature of thought.

Dynamics versus Computing

The recent surge in dynamical system studies has renewed a centuries-old debate about the nature of thought, argues Tim van Gelder, a philosopher at the University of Melbourne in Australia. The popular idea of the brain as a computer, he says, may not be as appropriate as describing the brain as a dynamical system. In a provocative article written for an unusual journal called *Behavioral and Brain Sciences*,[5] he points out that this debate in a cruder form was around long before computers even existed. Van Gelder says the seventeenth-century philosopher Thomas Hobbes viewed the mind in terms that today would be recognized as computational, for example. In the early eighteenth century, David Hume favored a Newtonian-like dynamical description of thought, with equations describing the association of ideas much like the way Newton's law of gravity described the attraction of planets to the sun. In Hume's view, "the dynamics of matter would be paralleled by a dynamics of mind," van Gelder says.[6]

Van Gelder is one of those philosophers who contend that the dynamical hypothesis is a valid option to replace the computational view that is currently dominant among cognitive scientists. "Cognitive systems may well be dynamical systems," he asserts.*

Well, there might be something to it. Presumably the atoms and molecules making up your brain's nerve cells are constrained to obey the laws of motion. If your brain is a dynamical system, its condition changes as the particles move around in whatever way the laws of motion dictate, determining what you will say or do. It doesn't sound entirely implausible.

But there is more to the story. The dynamical approach is one way of looking at a system, but not necessarily the only way. Some systems can be described in terms of computation just as easily as in terms of dynamics. Dynamics emphasizes one aspect of a system, the way it changes over time. Computing emphasizes something else— the arrangement of a system's parts at any given point in time. The arrangement of a brain's parts at one time can be regarded as a structure that represents information. To the computer-minded observer, the brain's operation is best described not as matter in motion, but as changes in structure. Or changes in information.

As time passes, the brain's structure changes in subtle ways to reflect changes in the brain's information content. Sensory signals (in-

---

*To be fair, van Gelder does not really claim that the computational approach is definitely wrong, just that the dynamical approach might be right. "Only sustained empirical investigation will determine the extent to which the dynamical hypothesis—as opposed to the computational hypothesis, or perhaps some other hypothesis entirely—captures the truth about cognition," he wrote in his article.

put) entering the brain change its pattern to form a new structure (output). If the brain's structure changes in a way that observes specific rules like the steps in a computer program, it makes perfect sense to regard the brain as a computer.

Basically, then, the debate boils down to which description of the brain works best: The computer approach, which analyzes thoughts in terms of producing outputs from inputs, obeying sets of rules, or algorithms; or the dynamical approach, which describes thoughts in terms of changing motions, governed by mathematical laws or equations.

Now if you're like me, you begin to wonder at this point why we should have to choose only one of these options to describe the nature of thought. Maybe the truth about thought resides not in one or the other of these possibilities, but in both. The brain is complex. Research into the nature of that complexity may show that explaining how the brain thinks requires more than one idea. In other words, the dynamical emphasis on change over time, and the computational emphasis on structure and algorithms, might both be needed to understand the complexities of thought.

That is precisely the argument of computer scientist Melanie Mitchell of the Santa Fe Institute in New Mexico, one of the brightest young thinkers in the computing and complexity business. She has given a lot of thought to how the brain is similar to computers, and different. During visits to Santa Fe I've talked with her at length about some of these issues and attended lectures where she has explored the connection between computing and thinking in compellingly insightful ways.

In a response to van Gelder's essay, Mitchell argued that both computational and dynamical ideas will be necessary to explain thought fully. The problem is reconciling the constant change of dynamics with the step-by-step processes in a computer. "What is needed," she says, "is a theory that incorporates both change and structure."[7]

She sees the seeds of such a theory in current work on theories of complexity—efforts attempting to explain how complex behavior arises from systems made up of large numbers of simple but interacting parts. Some of her recent work has shown intriguing connections between the dynamical and computational way of explaining things.

Simulations of a particular type of complex system show that processes that appear dynamical, like particles in motion, can perform the task of representing and conveying information. Dynamics and computation do not need to be enemies.

Similar views come from Mitchell's Santa Fe colleague, Jim Crutchfield. The essence of dynamical descriptions, he observes, is the importance of how a system changes in real time. In a complex system like the brain, those changes can lead to a rich variety of unpredictable behavior. Viewing the brain as a dynamical system offers a partial explanation for some of the general behaviors that brains exhibit. But dynamics alone cannot provide the brain's activity with usefulness and meaning, Crutchfield insists. Even if the brain can be described as a dynamical system, it nevertheless does represent and process information. The question is not whether the brain is dynamic or computes, he says, but rather how the dynamic system performs the computations.[8]

All this seems obvious to me, and in fact it illustrates one of the main points this book is trying to make, that different viewpoints illuminate different aspects of the same scientific phenomena. Sure, nature is full of dynamical processes, which is why Newton's laws of motion were so successful and Newton became so famous. But the clockwork determinism, force-guided motion of matter in the universe is not the only way to describe nature. Plenty of natural phenomena can be described quite effectively as examples of computation. Fireflies compute how to synchronize their flashes on hot summer nights, for example. Ants do all the engineering computations needed to create vast miniature transportation networks. Immune system cells gather information about germs and compute an antibody counterattack. And the brain can compute how to keep the image of your fingers stationary while you shake your head.

True, these "computers" of the natural world don't work the same way as a Macintosh or IBM clone. There is no central processor issuing orders to fireflies or to different parts of the brain. The individual bugs or brain cells operate almost on their own. Yet somehow the collective computational activity of simple units produces large-scale information processing.

Here, says Mitchell, is a clue to the mystery of life and intelligence and consciousness. The computational skills of life and intelligence don't conform to the standard principles of today's computing machines. It is hard to understand life and intelligence precisely because the principles underlying how living things compute remain a mystery.

"The brain's a very different kind of computer than the computers we're familiar with, if we want to call the brain a computer," she told me. "It has a very different kind of architecture. Instead of having a central controller, it has lots of relatively simple elements that are all connected in complicated ways. . . . We have a language for talking about computational systems. We don't have that language for computation in these other kinds of architectures."[9]

Learning to Compute

I quizzed Mitchell about these issues during the week in 1997 when she presented a series of lectures in Santa Fe, an annual event that draws a substantial audience from the community eager to hear insights from the edges of scientific investigation.[10] Mitchell's lectures explored the differences between brains and machines and tried to describe how "living computers" like the brain might be able to function without the central control that guides dead computers.

Basically, her approach involves simulating a device to see if it can learn to compute without any instructions. The "device" is one of those simulated networks with units, or cells, that can be either on or off. These networks, known as cellular automata, are very popular among people who study complex systems.

The common analogy for cellular automata is the game of Life, which doesn't even require a computer, merely a pencil to connect dots on a sheet of paper. The way the dots are connected must observe simple rules, which I unfortunately can never remember. So when I wrote about Mitchell's talks, I searched for another way of making her cellular automaton idea a little bit more concrete. I decided that what she was really talking about was a Christmas tree.

Just think of the units, or cells, as bulbs on a string of Christmas tree lights. These are lights with special circuitry causing any given bulb to turn on or off depending on what neighboring bulbs are doing. There is no central control—each bulb takes its orders only from its neighbors.[11]

In Mitchell's experiment, the string of bulbs would be designed to change its appearance every time it was plugged in. Let's say the power cord is plugged in once a day. Any given bulb would shine (or not) depending on what its two neighbor bulbs (and itself) were doing the previous day. Say bulb 27 was shining on Monday, but neighboring bulbs 26 and 28 were off. On Tuesday, bulb 27 might stay dark, while its neighbors turned on, depending on the rules of the system.

Mitchell and her colleagues try to simulate such systems to see whether coordinated behavior can arise without central control. By "coordinated behavior," she means some global aspect of the light string's appearance. The question is whether the system can learn to compute so that its global appearance represents the solution to some problem.

In the case of Christmas tree lights, the problem might be to have all the lights in the string be shining on Christmas Day. Or maybe to have all the lights turn off by New Year's. And in fact, Mitchell and colleagues have shown that the behavior of the whole string can be coordinated in that way, even if bulbs react only to their neighbors. One set of rules, for instance, might ultimately turn all the bulbs on by Christmas if more than half of the bulbs were on at the first plug-in, or turn all the bulbs off by New Year's if fewer than half of the bulbs were on initially. So even without central control, the string of lights could perform a computation leading to a desired behavior. Plotting the step-by-step progress of the light chain reveals a pattern of activity that looks very much like the plot of particle paths in a dynamical system. These "particles," Mitchell says, move in such a way as to allow the system to compress information about the environment and to communicate information throughout the system. In other words, dynamical systems can compute.

The trick, of course, is finding the local rules that will result in such global coordination. The secret to this, Mitchell confided, is computer sex. Don't be concerned, or disappointed, but this has nothing to do with Internet pornography. The sex is conducted entirely within the computer. The "mating partners" are the strings of numbers representing the computer code that tells the lights what rules to follow.

Mitchell tries different rules for telling the bulbs when to change and finds that some sets of rules force the lights closer to the desired behavior than others. She then takes the most successful sets of rules (encoded in strings of numbers) and allows them to mix and reproduce. It's the computer version of sex—parts of two sets of "parent" rules combine to make a new set of rules, the way two human parents contribute genes to their child. After many generations of allowing the most successful computer rules to mate, some sets emerge with the ability to solve the desired problem.

In essence, then, the Christmas-bulb computer produces its own program. It seems to Mitchell that similar computation happens in biology. Just as individual lightbulbs can learn to cooperate to turn on the whole string, individual ants can cooperate to construct an intricate colony with a superhighway tunnel system, for example. It is not too great a stretch to think that something similar goes on in the brain when neurons develop coordinated communication necessary for thinking. Mitchell hopes that computation research of the sort she is pursuing will help explain how sophisticated behavior like tunnel construction or thinking can emerge when simple things, like ants or neurons, process information.

Ultimately, she suspects that this understanding will show the way to a deeper appreciation of the relationship between machinery and life. Computer versions of "life," or intelligence, may narrow the gap between the inanimate realm of machines and the world of biology. Life may someday be understood in mechanistic terms—as a special kind of machine, she believes—as ideas about what a machine is, and about what life is, change. After all, she points out, ideas of what a machine is have changed dramatically since Isaac Newton's day, when machines were things like wedges, levers, and screws. Computers are machines of a very different sort.

"I think our intuition about what 'mechanistic' means will continue to change," Mitchell said during her lectures. "As computers get more and more sophisticated, our notions of what a machine is will change. At the same time, our notion of life gets more and more mechanistic." So as the notion of machine becomes more biological, the notion of biology will become more machinelike. "Somewhere in the middle they'll meet," she said. "I don't know where that will be."

Well, perhaps it will be in Georgia.

Computing with Chaos

At the Georgia Institute of Technology, William Ditto has been studying dynamical systems and life for years, investigating such issues as how heartbeat dynamics differ in diseased and healthy hearts. Lately he has turned from the heart to the mind, devising a specific example of a dynamical system—a chaotic dynamical system at that—that can perform general purpose computations.

Ditto's idea for chaotic computing shows that certain kinds of dynamical systems can, in principle, compute in much the same way as ordinary computers. In fact, he suggests that chaotic dynamical systems could become an entirely new genre of computing, in a class of its own comparable to ordinary digital computing, quantum computing, or DNA computing. With his colleague Sudeshna Sinha of the Institute of Mathematical Sciences in Madras (now known as Chennai), India, Ditto has simulated a network of fast chaotic lasers connected to each other and tuned in such a way that the chaos itself would compute answers to math problems. Numbers could be represented in various ways—by the timing or the brightness of the laser pulses, for example. Once the input numbers are represented in the system, one laser's pulse would affect the pulsing of its neighbors. A chain reaction of pulses would follow. At some point, a cascade of pulses (the lasers' equivalent of the sandpile avalanche) would flow to the edge of the network—to its interface with the outside world—to deliver the answer to the problem.

So far, Ditto and Sinha's simulations (on an old-fashioned computer) have shown that a chaotic system can in principle conduct logic operations to transform input into output the way the logic gates of standard computers do. The paper, published in 1998 in *Physical Review Letters*, further describes how such a system could add and multiply and even solve more complicated problems. And you wouldn't necessarily need to use lasers—in principle, any chaotic system could be used to compute. "Hell, you could use coupled dripping water faucets if you had a reason to," Ditto says.[12] But fast lasers are the more logical choice for rapid computing.

Ditto's main interest in this isn't putting CompUSA out of business. He wants to figure out how to connect computer chips to real nerve cells. To his wife, Ditto says, this sounds like he is trying to build a Terminator. But Ditto sees it as a way to find out if a machine can really simulate human intelligence. "What we are really trying to do," he says, "is build a computer that behaves more like a brain."[13]

Conscious Computing

The goal of building a computer that behaves like the brain might be easier to achieve if anyone knew just how the brain does behave. For most of the past century that was widely considered to be a forbidden question. You could observe how people behaved, but not what was going on inside the brain that correlated with that behavior. In particular, the notion of studying consciousness was generally discouraged. But in the last decade or so many of the old prohibitions against studying consciousness have lost their grip on science. For one thing, brain imaging technologies have made it possible to peer inside the skull without a scalpel. In response, philosophers and scientists alike have consumed a lot of ink in attempting to explain consciousness.

"Fifteen years ago practically nobody was working on it and not very much was being written about it, even by philosophers at that time," Francis Crick pointed out when I talked to him in 1998. "That has changed. But in some ways it might be said that it has changed for the worse, that there are too many people talking about it and not much doing."[14] In one of Crick and Koch's papers, they summarize the situation by quoting a breakfast menu from a Pasadena hotel: "When all is said and done, more is said than done."

Still, if the brain really accomplishes its magic by performing computations, then it ought to be possible to explain, step by step, how the brain works. And it even ought to be possible to explain consciousness. Curiously, though, some recent efforts to explain consciousness explicitly reject the idea that the brain is a computer. Roger Penrose, you'll remember, argues vigorously that consciousness cannot be mere computation. He draws on fundamental elements of computer science, neuroscience, and even quantum physics in reaching rather grand conclusions about the noncomputational nature of human thought and the physical mechanisms of consciousness.

Penrose believes that consciousness has something to do with quantum effects in microtubules, structures that form a sort of scaffolding within brain cells (and other cells) used for transporting molecules around. Since his books have appeared, I've discussed his conclusions with prominent computer scientists, neuroscientists, and quantum physicists and have yet to find anyone from these specialties who thinks Penrose is on target.[15] "It would be remarkable if Penrose were right," was Francis Crick's only response when I asked him about Penrose's ideas.

In any case, there doesn't seem to be the slightest shred of evidence for any of Penrose's conclusions.[16] I think his main problem, and the problem of many other attempts to explain consciousness, is lack of connection to what goes on in real brains. Crick agrees. "There's far too much talk and very few people doing experiments," he said. "We think that will change. Not that I want lots of people to do experiments, but I would be happier to see a few more doing it."[17]

Crick and his colleague Koch are content to develop theories about how consciousness works, focusing on the one aspect of consciousness where progress has been greatest—visual awareness. They argue that when a brain is conscious (in this case, visually conscious or aware), something must be going on in the brain that makes such awareness possible. And whatever is going on must involve the activity of some specific nerve cells. Therefore, as Crick and Koch see it, the issue is finding those neural correlates of consciousness.

Neuroscientists have long known that the brain's "vision center" is at the back of the brain in the occipital lobe. At least that is where the optic nerve sends signals from the retina. (Actually, the optic nerve sends signals to a relay station called the lateral geniculate nucleus; the signals are then sent on to the visual cortex at the back of the brain.) That occipital part of the cortex (the brain's rumpled outer layer) contains just the first stage of visual processing, however. After all, the signals initially entering the brain are just wild splashes of light and color. Several stages of processing (as though the brain were a sophisticated darkroom) are required to generate an image of a sensible scene to the mind's eye.

There is no point going into the unending complications of visual processing in the brain. But Crick and Koch offer intricate analyses to argue that visual awareness critically depends on brain activity not in the back of the brain, but in its frontal regions.

I've talked to neuroscientists who don't think Crick and Koch are on the right track, and certainly the jury is still out. But in this case there is some real evidence favoring their ideas about where the consciousness action is. The most impressive evidence, I think, comes from the bizarre phenomenon known as blindsight.

Lawrence Weiskrantz, a neuropsychologist at Oxford University in England, is a pioneer in the use of brain scans to study blindsight. It occurs in some people with head injuries to the back of the brain, where the main vision-processing brain tissues reside. Damage to the brain's outer layer, or cortex, on the left-back of the head can leave the victim totally blind in the right side of the field of view, as is the case with a famous patient known as G.Y.

G.Y. is ordinarily unaware—that is, he's not conscious—of light or color in his blind field. But somehow his brain can still get the message. Weiskrantz's experiments have shown that even when G.Y. cannot see a spot of light, he can correctly guess which direction it is moving. In other words, the brain receives information that its owner isn't conscious of.[18]

Now if the experiment is changed a little—the light is moved more rapidly, for example—G.Y. will report awareness that something is going on, though he still can't describe what. So in this case G.Y.'s brain activity reflects an awareness of some sort. In other words, something is happening in the brain that is related to G.Y.'s being conscious of the light's motion.

G.Y. therefore offers a unique opportunity to find out what specific brain activity is correlated with consciousness. When most people are conscious of something, too much other brain activity is going on to tell which part of it is related only to consciousness itself. But with G.Y., experimenters can scan his brain when he's "aware" of the light and when he's not—the only difference is the awareness, or consciousness.

Sure enough, Weiskrantz and colleagues have found that the brain's pattern of activity changes depending on whether G.Y. is aware of the light or not. In fact, various brain regions show different activity levels when G.Y. is aware of the motion. And one of those areas is in the frontal cortex, precisely the region that Crick and Koch have predicted. But too much is going on to say that Crick and Koch are surely right. A more important implication of the blindsight study in G.Y. is that brain activity connected to consciousness can be isolated. The reasonable conclusion is that understanding consciousness is a legitimate goal for scientists to pursue experimentally, and that it is not necessarily an impossible undertaking.

Where such research will lead is still hard to say. Whether the computational approach will be the most fruitful guide for figuring out the brain is still not entirely clear, either. But it seems inconceivable to me that describing the best information processor biology has ever invented will not require insight into information processing.

A Complex Explanation

Somehow or another, any explanation of consciousness, and more generally the brain's information-processing skills, will have to cope with the brain's complexity. The brain's multiple information-processing abilities emerge from trillions of connections among billions of nerve cells. Ultimately, explaining the brain will no doubt depend on explaining how the brain gets to be so complicated. That, of course, is a huge field of research, the study of neural development. And lurking beneath the surface of that field is the contentious question of whether behavior is genetically programmed or orchestrated by the environment.

On the gene-environment issue (commonly known as the nature-nurture debate), most scientists concede that behavior must result from an interplay between genes and environment. The real question is how. There are lots of ideas. One comes from the neurobiologist Michael Gazzaniga, and I like it because it fits so perfectly into the information-processing picture. He suggests that the solution lies in comparing the brain to the immune system. In his popular book *Nature's Mind*, Gazzaniga notes that disease-fighting antibodies seem able to attack any foreign invader, or antigen, regardless of the shape of the invading molecule. "The antigen . . . selects out the most appropriate antibody molecule, which either preexisted in the body or evolved from strict genetically driven processes," he writes.[19] "Just as each of us has a unique set of antibodies undoubtedly controlled by DNA, we may also have a unique set of neural nets, which enable us to have different capacities."[20]

In this view the brain is neither hardwired from birth nor a blank circuit board waiting for the environment to etch neural pathways. The brain is a bundle of possibilities, of nerve cell circuits carrying various capabilities, with only some selected during development through "give-and-take" with the environment. Children learn language quickly because a capacity for language has been built into the brain, but which language they learn depends on whatever is spoken in their environment.

In other words, the brain is shaped by information. Just as the evolution of species can be viewed as information processing, and the immune system's disease-fighting strategy can be regarded as computing effective counterattacks, the development of the brain is an exercise in processing information. The developing brain takes in information and responds by creating a structure that can process more information. Instead of natural selection of species or antibodies, there is a natural selection of nerve-cell circuitry.

The idea of nerve-cell circuit selection as the basis of consciousness has been championed by a neuroscientist who started out (and won a Nobel prize) as an immunologist, Gerald Edelman. In fact, Edelman, of the Neurosciences Institute in San Diego, coined the term "neural Darwinism" to describe the development of a complex brain through selection of neural features by the environment. The unit of selection analogous to the species in Darwinian evolution is an interconnected group of nerve cells (or neurons) called a "neuronal group."

Edelman therefore calls his theory of the brain "neuronal group selection." In his view this process governs how brain anatomy develops, how experience selects patterns of responses from this anatomy, and how signaling between neuronal groups gives rise to behaviors that influence an organism's ability to survive. Neuronal group selection, he argues, can explain the brain's incredible complexity and abilities, including consciousness itself.

In Edelman's theory, selection happens in two major ways. One occurs during the time the prenatal and infant brain is growing, when interaction with the environment guides the development of neuronal groups and their connections. Then, later on in life, experiences alter the strength of connections between neuronal groups. So any individual's mental capabilities will therefore depend on the pre-existing features of the brain produced by genes (the result of a long evolutionary history) and the specific environmental interactions that select some of those features during life.

Edelman is regarded rather contemptuously by many scientists—more for reasons of personality than science, I gather—but that is of no concern to me. I have never interviewed him. But I've read many of his papers and one of his books and tried to convey in the newspaper what he was talking about.[21] As I managed to grasp it, the key to his explanation of the brain lies in understanding the flow of specific feedback signals back and forth between different sets of neuronal groups. This process, which he calls re-entrant signaling (or just re-entry), coordinates the responses of various neuronal groups to inputs from the senses.

In his book *Bright Air, Brilliant Fire*, Edelman applied these basic ideas to consciousness itself. He starts by distinguishing two forms of consciousness. One is the here-and-now "primary consciousness," an awareness of what's going on at the moment, but without a sense of past or future. Household pets appear to have primary consciousness. A cat has some idea of what's happening now but never makes long-term vacation plans, for example.

The other form is the higher-order consciousness possessed by humans. It involves a sense of self, an ability for a person to think about the past and future and about his or her acts and feelings. The difference between cat consciousness and people consciousness, Edelman explains, is that "we are conscious of being conscious."[22]

Edelman's explanation of consciousness is elaborate, subtle, and complicated, too complicated to condense successfully into a few paragraphs. But you can get the general idea if you think about the brain as built around value centers. The value centers are not hardware stores, but brain areas that tell animals what's good for them.

Value centers evolved for some pretty basic purposes, such as providing a built-in desire to eat when hungry or sleep when tired. Because they go way back in evolution, the basic value centers are found in the brainstem and other nearby brain areas, the "older" parts of the brain.

These value centers are basically concerned with events inside the body. A second nervous system copes with the world outside. That system is based in the brain's convoluted outer covering, the cerebral cortex, the "newer" part of the brain from an evolutionary standpoint. The cortex receives signals from the senses.

In Edelman's view, the modern conscious brain is the result of the merger of these two nervous systems—the value system, monitoring the inside of the body, and the sensory system, tuned into the world outside. To do these jobs the brain's nerve cells are organized into committees (Edelman's neuronal groups). The nerve cells in a committee are linked and work as a unit on a particular task.

The neuronal committees have to communicate with the rest of the brain too, of course—they're all wired into a "Brain Wide Web" of nerve fibers that send signals back and forth between committees, making sure that everybody shares the messages from the value centers and from the senses. Consciousness, at its most basic, is all about piecing together the information flowing among all these committees and making sense of it all.

It's not easy, of course. The nerve cell committees all have specialized tasks. Some are in charge of noticing colors, some recognize shapes, others respond to motion. They all signal each other so that the brain can match up the various committee reports and draw sensible conclusions. For example, a certain combination of shape, color, and motion would match up to tell the brain that "a baseball is headed this way." The nerve cell committees in charge of responding should then direct the body's muscles to perform an appropriate behavior—perhaps reaching up to catch the ball.

How does all this work? Edelman thinks that the nerve cell committees organize experiences into categories—a flying baseball might be categorized as danger; a falling apple might be categorized as food. These categories are naturally shaped by the input from the value centers. What Edelman calls "primary consciousness" draws on experience and values to categorize the many signals arriving from the environment. Combining perceptions with values in that way can guide an animal's immediate behavior.

Primary consciousness alone, however, cannot recall symbols and their meanings. It's not enough just to categorize input from the outside world—the conscious brain must be able to categorize its own activities as well. That, in Edelman's picture, is the key to forming concepts. Concept formation is the job of nerve cell committees that pay attention not to the outside world, but to what other neurons are doing. As you can see, the brain is a very complicated place.

By keeping track of the brain's activities, these committees can make memories of which brain actions turned out to be good ideas—beneficial for meeting the body's needs. Ultimately these committees are also responsible for memory of concepts, social communication, and language. Those abilities give humans a "higher" form of consciousness, not tied to the here-and-now. People can think about the past and the future, and even be aware of their own awareness—thanks to all the back-and-forth signaling between value centers, sensory committees, and concept committees.

As it stands, Edelman's approach offers an intriguing set of ideas possibly complex enough to reflect what really goes on in the brain.[23] But what does it have to do with information? It turns out that the sharing of information among neuronal groups can be described in terms of Shannon's information theory. Edelman and his colleagues Giulio Tononi and Olaf Sporns have worked out rather elaborate mathematics to measure the complexity of signal matching in the brain.

The idea is that the brain's intrinsic complexity—its pattern of connections—is only part of the story. There's another aspect of complexity that Edelman and colleagues call the "matching" complexity. It depends on the amount of information exchanged through those connections among the brain's nerve cell committees. "Matching complexity is a measure of how much information is exchanged, on average, in the brain," Edelman's collaborator Giulio Tononi explained at a neuroscience meeting I attended in 1998.[24]

Tononi, Sporns, and Edelman say that their analysis of information exchange makes sense of the back-and-forth, re-entrant signaling view of how the brain works. In particular, different amounts of information sharing among those neuronal committees can explain how the brain accomplishes the competing goals of specialization and cooperation. Some committees have to specialize by responding only to specific categories of sensory input. Yet the whole brain has to cooperate so that the categories of input can be associated with values and appropriate behaviors. The information-theory analysis shows how these competing goals can be achieved by different amounts of information sharing between subsets of neurons.

In 1998, Tononi and Edelman elaborated on this view in an intriguing review article published in *Science*. They concluded that consciousness is maintained by various sets of neurons that measure high on the diversity (information-sharing) scale and form clusters with strong, rapid feedback. Tononi and Edelman call a cluster of neuron groups that meet these requirements the "dynamic core"— reflecting its role in the unity (core) of consciousness plus its ability to change rapidly (dynamic).

"Its participating neuronal groups are much more strongly interactive among themselves than with the rest of the brain," Tononi and Edelman wrote. "The dynamic core must also have high complexity: its global activity patterns must be selected within less than a second out of a very large repertoire."[25]

I have no idea whether this view of the brain will turn out to capture the essence of how intangible states of mind are generated by the matter in the brain. Dozens of other deep thinkers have proposed descriptions of consciousness that are quite different. And the Edelman approach is just one example of how quantifying information might be used to explain some of the mysteries of the brain's biology. There have been countless other books, symposia and conferences, journal articles, and Web pages proposing elaborate explanations of how information is processed in the brain and how it all relates to consciousness.

One thing is sure—most of that stuff is all wrong. But some small portion of it will someday turn out to be prescient. The trouble is, there is no sure way to know now which part that is.

In any event, it seems pretty likely that any ultimate explanation of the brain's activity will somehow have to encompass its information aspects. The brain is no mere dynamical system like the solar system or the weather. The brain gathers and uses information. In other words, the brain is an information gathering and utilizing system—an IGUS.

Chapter 8

IGUSes

The Church-Turing principle states that the laws of physics are such that one specific machine can, in principle, simulate any physical system. . . . The laws of physics must be consistent with the existence of human thought and life itself. (DNA is an information-storage medium, and biological evolution, which generates new genes, is a form of information processing. Genes could not evolve to fill ecological niches if the problems set by the physics of sunlight, water and gravity were not soluble by computations performed by quite different physical systems, such as long-chain carbon molecules.) Thus the Church-Turing principle is a sort of anthropic principle without the explicit anthropic reference.

—DAVID DEUTSCH,
"Quantum Computation"

In his last public appearance, at a physics class in Princeton, Albert Einstein declared one last time his displeasure with the prevailing philosophy of physics.

"If a person, such as a mouse, looks at the universe, does that change the state of the universe in any way?" he asked.[1] To Einstein, it made no sense to suppose that a person could alter reality by observing it, any more than a mouse could.

But some physicists have suggested otherwise. They've pointed out that many of the mysteries of quantum mechanics imply the need for an observer to play an essential role in shaping reality, at least on the subatomic level. If observers have this power on a small scale, why not on large scales? Why not the entire universe? Maybe not a single mouse. Maybe not a single human. But somehow, in some mysterious way, perhaps the existence of the universe has something to do with the existence of life.

After all, certain properties of subatomic particles, nuclear processes in the interior of stars, and the cosmic forces that shape the universe as a whole all seem fine-tuned to permit life to exist. As the physicist Freeman Dyson once expressed it, to say life arose by chance is merely an expression of ignorance. "The more I examine the universe and study the details of its architecture, the more evidence I find that the universe in some sense must have known that we were coming."[2] Either these fine-tunings are remarkable coincidences, many scientists have argued, or there is some deep connection between life and the most fundamental features of physical reality. Some, including John Wheeler, have speculated that the universe is built in such a way so as to make life inevitable. "Is the machinery of the universe so set up, and from the very beginning, that it is guaranteed to produce intelligent life at some long-distant point in its history-to-be?" he asks. "Perhaps."[3]

The coincidences of cosmology and quandaries of quantum mechanics have impelled a number of serious scientists to entertain the notion that, in some incompletely understood way, the universe does in fact require the existence of life. That view has been elaborated in various guises under the label "anthropic principle." Some of the ideas behind the anthropic principle have been discussed by philosophers for centuries. But significant interest among scientists began to grow only in the early 1970s, when the British physicist Brandon Carter pointed out that small changes in many of the basic numerical constants of physics would render life impossible.*

---

*The modern interest in the anthropic principle can be traced to an observation made by the American physicist Robert Dicke in the 1950s. Much earlier (in 1937) the British physicist Paul Dirac noted that certain "large numbers" (such as the age of the universe and the ratio of the electric force to gravitational force between a proton and electron) seemed about equal. To explain

The strengths of nature's forces are determined by the values of a handful of fundamental physical constants, including the speed of light, the electric charge of an electron, and Newton's gravitational constant. These constants capture the essence of physical reality, describing what nature is like in a quantitative and precise way. Even very small changes in these constants would change the strength of nature's forces in such a way that life would be impossible, Carter and other anthropic-minded scientists pointed out.

If the force holding the atomic nucleus together were about 10 percent weaker, for example, no element heavier than hydrogen could have been created; the carbon and other elements necessary for life as we know it would not exist. If the nuclear force were about 4 percent greater, all the hydrogen would have been converted into helium when the universe was very young. No stars would have formed to cook hydrogen into the more complex elements necessary for life.[4] If the electromagnetic force were somewhat stronger (or the nuclear force somewhat weaker), elements like carbon would have been unstable; even if they formed, they would have broken apart before life became possible. Changes in other constants would alter the nature of stars, making them all either too hot or too cool to support a planet with life. Even the size of the universe, which seems a lot bigger than necessary to accommodate an inhabited planet, must be what it is to permit life. If the universe were a lot smaller (say, big enough for only one galaxy instead of billions), it would have expanded and then collapsed in about a year. There would be no time for life to originate and evolve. So it appears that we live in a "Goldilocks" universe—its size, its age, and the strengths of its forces all came out "just right" for life.

Carter translated these "coincidences" into the "weak" version of

---

such coincidences, Dirac proposed that the strength of gravity varied with the age of the universe. Dicke, however, argued that the near equivalence of such numbers really was just a coincidence. He could "explain" it by pointing out that the universe has to be about the age it is before people would be around to raise such issues. In a universe much younger, stars would not yet have produced the carbon and other heavy elements needed for life. In a universe much older, no stars would remain shining to support life. So Dirac's "large-number coincidence," in Dicke's view, could be explained by anthropic reasoning.

the anthropic principle—that the universe must have properties that make life possible. That does not seem especially shocking. It would appear to be automatically true if anybody is around to write it down. But it is still curious that even slight changes in the universe's essence would make life impossible. Perhaps the universe was trying to tell us something.

During the 1980s, considerable attention was given to these ideas, and the anthropic principle was embraced by some physicists as an almost religious view of the connection between the universe and life. Most favored the weak form, but a few adopted the stronger belief that the universe was not only hospitable to life, but required the existence of life. The peak of this anthropic fever came in the mid-1980s with the nearly concurrent publication of two very different books. One was *Roger's Version,* by the novelist John Updike. It chronicled the adventures of a young theology student who explored anthropic principle physics in his search for God. The other, by John Barrow (a British cosmologist) and Frank Tipler (an American physicist), was a compendium of anthropic evidence called *The Anthropic Cosmological Principle*. It cataloged evidence favoring the anthropic view from a variety of fields of study and identified three distinct versions of "the" anthropic principle.

The weak anthropic principle:
"The observed values of all physical and cosmological quantities are not equally probable but they take on values restricted by the requirement that there exist sites where carbon-based life can evolve and by the requirement that the Universe be old enough for it to have already done so."

The strong anthropic principle:
"The Universe must have those properties which allow life to develop within it at some stage in its history."

The final anthropic principle:
"Intelligent information-processing must come into existence in the Universe, and, once it comes into existence, it will never die out."[5]

The first two versions reflected the authors' assessment of the literature in the field; the third was their own creation, the organizing principle for their book. The three principles invited the abbreviations WAP, SAP, and FAP. It wasn't long before Martin Gardner, the *Scientific American* columnist known for his disdain for pseudoscience, suggested that FAP might be better named the completely ridiculous anthropic principle, which could be abbreviated CRAP.[6]

Indeed, most scientists I've encountered think that Tipler and Barrow's speculation went way too far. In preparing a series of articles on the anthropic principle in 1987, I asked several cosmologists and astrophysicists what they thought of it. Most were happy enough with the weak form. "It's a perfectly legitimate scientific inquiry," said William Press of the Harvard-Smithsonian Center for Astrophysics. "We're trying to discern the nature of the universe by direct scientific observation. One of those observations is the fact that we're here, we're made up of carbon organic chemistry, we live around a star with certain characteristics, and we're certainly entitled to take those observational facts and see what conclusions we can therefore draw about the way that the universe must have been."[7] J. Richard Bond pointed out that the anthropic principle can be useful to physicists in working out theories of the early stages of the universe. Whatever happened then had to lead eventually to a universe where life could begin and evolve. Of course, both Bond and Press warned against taking the idea too far. "I think it is very interesting to probe the coincidences required for life," said Bond. "The danger is to go off into the realm of philosophy without data to keep you honest."[8]

Press objected vigorously, though, to the idea that a universe without life cannot exist. "I get off that train before the end of the line," he said. Tipler and Barrow's book goes too far in contending "that the actual evolution of the universe may have something to do with our being here rather than the other way around. That's certainly not a part of the book that I subscribe to."

Other scientists were not so kind in their assessment. The late Heinz Pagels, a physicist who wrote several popular books before dying in a mountain climbing accident in 1988, also deplored the anthropic principle bandwagon. In a talk at a scientific meeting, he said:

"The anthropic principle is a sham, an intellectual illusion that has nothing to do with empirical science. The anthropic principle promotes a kind of cosmic narcissism regarding the role of life and consciousness in the universe, and while it's fun to speculate about the cosmos, the serious work is meanwhile not getting done. The anthropic principle is a lazy person's approach to science."[9]

Criticisms such as those from Gardner and Pagels were perhaps excessively harsh. Plenty of serious work is still managing to get done. And there have been many serious uses of the anthropic principle, in its weak form at least.[10] Still, the anthropic principle somehow seems unsatisfactory. The weak form does not appear to offer a very deep insight into the existence of the universe—it's more descriptive than explanatory. And the strong form remains rather hard to believe. If observers generated existence, what generated observers to begin with? Somehow the anthropic viewpoint did not make the mysteries of the observer's role in quantum mechanics any easier to solve.

As anthropic enthusiasm has faded, however, a new view of observers is emerging from a perspective combining quantum physics and information processing. If information is a fundamental aspect of existence, as basic as matter or energy, then it makes sense to think that observers have some role to play, for obtaining information is the very essence of "observation." So perhaps the information-processing perspective can provide a new, more fruitful way to link observers to reality. Rather than some sort of magicians pulling reality rabbits out of a quantum hat, observers can more fruitfully be regarded as entities that acquire and process information. This point of view gives observers their own acronym—IGUS—for information gathering and using (or utilizing) systems. It's a view that opens the way to a whole new understanding of the relationship between observers and reality, the nature of science, and even the evolution of life.

Complexity

My first glimpse of how these ideas come together came on a cold Tuesday in Chicago. It was February 1992, at a meeting of the American Association for the Advancement of Science, in a lecture by Murray Gell-Mann, the inventor of the concept of quarks. His was the last talk in a session on the last afternoon of the meeting, and only a few dozen people stayed around to listen. I congratulated myself on being smart enough to be one of them. In a little more than an hour, Gell-Mann managed to weave all the important aspects of science together—life, the universe, and almost everything.

Gell-Mann is one of those scientists whose legend is larger than life. He is known above all for being very smart (his name always turns up on those meaningless "smartest people in the world" lists). And of course he is very smart, and is also very critical of people who aren't, a trait that earns him a lot of criticism in return. Some people just don't get along with him. But he is not the only scientist who refuses to suffer fools, and he does have a gracious and charming side. It's also refreshing to talk to someone who speaks frankly; often scientists express a criticism so politely that it's hard for a journalist to recognize it as critical. I prefer straight-shooters—especially when they have a secure intellectual command of the scope of science. And nobody's scope of knowledge exceeds Murray Gell-Mann's, as his talk in Chicago demonstrated. He seemed to fit everything about science neatly into place, all attached to the common thread of understanding complexity.

Complexity is like a lot of other ideas in physics; it has a common, intuitive meaning and it has a specific, technical meaning. At least it ought to have a specific technical meaning if it is to be useful. And in the last decade or two some very bright people have spent a lot of time trying to make the notion of complexity specific and precise. In the process they've come up with not one, but many technical meanings for complexity. But almost all these approaches naturally converge with ideas about information and information processing, and that inevitably brings observers into the picture. Observers, in turn, are themselves complex. So understanding complexity requires understanding observers, and understanding observers depends on understanding complexity.

The first step toward understanding complexity is defining the rules of the game, Gell-Mann pointed out in his lecture. And the first rule is to define the level of detail to be considered. Without specifying a level of detail, questions about complexity cannot be unambiguously answered. As an example, Gell-Mann cited a common debate among ecologists: whether a complex system is more resilient than a simple system. One school of thought suggests that complexity is beneficial for a system, providing it with better survival skills, so to speak. An opposing school holds that a simpler system would be more stable against disruption by an outside force—a fire, say, or some natural disaster or man-made catastrophe.

Gell-Mann observed that advocates for simplicity appeared to be winning the debate, at the moment. But it wasn't clear to him that the notions of simplicity and complexity were being adequately defined. He compared a simple mountain valley to a tropical forest. The valley probably contained only a few species of trees, probably fewer than 10, while the forest is populated by hundreds of tree species.

"But why only trees?" Gell-Mann asked. Perhaps counting insects would be a better measure of complexity of the system. Or perhaps it would be better to quantify various interactions—between predators and prey, pollinators and pollinated. His point was simply that a level of detail must be specified before any consideration of complexity can be fruitful. In physics, this notion is known as coarse graining—establishing the degree of detail to be considered.

Next, Gell-Mann emphasized, it is important to specify the sort of language that will be used to describe a system. And the description must be such that a "distant correspondent" will be able to understand and interpret it. (You shouldn't be allowed to describe a very complex system just by pointing to it, for example.) Given an agreed-on language and a specified coarse graining, the complexity of a system then ought to have something to do with the length of the message needed to describe it. Gell-Mann chose to define the "ideal complexity" as just that—the length of the shortest message that can fully describe the system to someone not present.

Measuring the length of a message is easy enough in the age of computers—all you need to do is put the message in a computer and count the number of bits of information the message takes up in the computer's memory.[11] Or to make it even easier, just suppose that the agreed-on language was binary numbers, so the message would be expressed as a string of 1s and 0s, or bits, to begin with.

Defining complexity in this way isn't so simple, though, because computers can play interesting tricks with strings of bits. For example, you could program your computer to print out the message describing a system, and the program might be much shorter than the message itself. In that case the "shortest message" would not be the message you started with, but the program that can print the message. To give an extreme example, suppose the string of bits looked like this:

```
11111111111111111111111111111111111111111111111111
11111111111111111111111111111111111111111111111111
11110000000000000000000000000000000000000000000000
00000000000000000000000000000000000000000000000000
00000000
```

That's 200 bits. But you could condense that bit string into a very short computer program:

For $x = 1$ to 100, Print 1
For $x = 101$ to 200, Print 0
Stop

Since the program contains all the information it needs to describe the system, the program would be the shortest form of the message. So, Gell-Mann observed, it makes sense to define the "ideal" complexity of a system in terms of the computer program needed to produce the description of that system. Specifically, the ideal complexity would be the length of the shortest computer program that prints out the string of bits describing the system, and then stops. (You also need to specify the computer the program will be run on.)

As it turns out, the ability to shorten the description—compressing the message—is the key to understanding the nature of observers. There's nothing mysterious—or mystical—about the role of observers in the universe. Observers are simply complex systems that have the ability to compress information about their environment.

"If you're a serious theoretical scientist, you don't believe there's anything special about people in the universe," Gell-Mann asserted. "There must be all sorts of things in the universe that can interpret their surroundings and compress data the way we can."[12]

Only a rather complex system could interpret its environment and compress data. But not all complex systems can compress data. The ones that do are special. Gell-Mann and other scientists on the frontiers of complexity research call the special ones "complex adaptive systems"—a kind of complex system defined by their ability to adapt and evolve.

People are complex adaptive systems. So are amoebas. And goldfish. And living things in general. For that matter, many products of living things are also complex adaptive systems in themselves, such as human language, the world economy, and the entire scientific enterprise.

Not all complex systems are adaptive, Gell-Mann noted. Thunderstorms in the atmosphere or whirlpools in a rushing stream are complex, but not adaptive. They don't compress information. An amoeba, on the other hand, compresses within its genes all the information needed to make another amoeba.

Reproduction is not the defining feature of complex adaptive systems, however (even the eddies in a whirlpool can produce new eddies). The key common feature of complex adaptive systems is the ability to process and condense information to generate what Gell-Mann calls a schema, a sort of guide to interacting with the environment. A schema encodes a system's experience into principles that summarize the regularities of nature.

A schema provides an internal picture, or model, of what nature is like and how it behaves. Identifying regularities in the environment is the essential feature of making a schema. A simple example of a regularity is "what goes up must come down." Lots of different things come down after going up—baseballs, footballs, bricks, red bricks, cream-colored bricks, rabbits, and rocks. A schema would be worthless if it had to encode for every individual possibility. By encoding only regularities, a schema can contain a condensed description of complex things, allowing a complex adaptive system to adapt to the regularities in its environment.

Regularity and Randomness

Of course, not all descriptions of a complex system can be compressed. Look at this string, for example:

0001001100011111101001110100011110000001111000000
1011

It is hard to detect any pattern that would make it feasible to write a short program to print this string. Probably the shortest program would be just:

Print:
0001001100011111101001110100011110000001111000000
1011

Information strings that cannot be compressed are called "random." The above random string was produced by flipping a coin 52 times. But if you don't know that a string of bits was produced randomly, you can never be completely sure that it is truly random. It might contain some subtle pattern that would make it possible to compress the message. And in fact, complete randomness is rare in nature. Most of the messages describing reality can be compressed into shorter forms because most things exhibit some degree of regularity, or pattern. The key to constructing a schema is to compress the regularities of nature and forget the random stuff.

"It's crucial to realize that the system's record of experience is not just a compression of the whole of the experience," Gell-Mann emphasized. "The crucial feature is selecting what appears to be regular and discarding . . . what appears to be random."[13]

A complex adaptive system has to live by the rules it creates in condensing information about nature's regularities. Organisms with good schema—schema that encode nature's regularities accurately—can make better predictions about the results of their actions and therefore are more likely to survive.

A similar process underlies the scientific enterprise itself, Gell-Mann observed. Scientists observe regularities in the world and encode information about those regularities into a schema, or theory. The theory's equations are then applied to the random situations encountered in nature (the "boundary conditions," in physics jargon). Over time, theories evolve to better adapt to what really happens, much in the way living organisms evolve because different genes respond differently when encountering randomness in the environment.

"Scientific theory reduces a large body of data, actually indefinitely large, to some simple equations plus boundary conditions," Gell-Mann explained. "Now the boundary conditions are not simple. And the boundary conditions may well be random. But there are a lot fewer boundary conditions than there are situations."[14]

In essence, the equations sum up all you know about how nature works, Gell-Mann said. Then you can describe a system just by specifying the particular conditions to apply the equations to. The "acme of simplicity," Gell-Mann maintained, would be the compression of the fundamental laws of physics into one theory that describes all of nature's fundamental particles and forces. Then all you'd need would be to apply the math of that theory to the ultimate boundary condition—the condition of the universe at the instant of the big bang—to have a complete description of the fundamental principles underlying all of physical science.[15]

Encompassing as it did the fundamental principles underlying the entire universe, Gell-Mann's talk was what I like to call information-rich. It was full of substance and insight. But that also means there was a lot to ingest. In particular, it became clear to me that there was a deeper (and more complex) connection between the ideas of information and complexity than I had previously realized.

For one thing, as Gell-Mann pointed out, his "ideal complexity"—the length of the shortest computer program describing a complex system—was not necessarily the most useful measure of the system in question. After all, some very complex messages are complex only because they are random, and random information cannot be compressed. It might be a better idea, he said, to regard the length of a schema needed to describe a system as the more important measure, thereby emphasizing the regularity and disregarding the randomness. But the length of the schema isn't all there is to consider. Some credit is due for the amount of time and effort needed to produce the schema. Scientists spend a lot of time identifying regularities in nature and devising equations to represent those regularities, for example. Great artists put a lot of work into painting a picture rich in complexity, and evolution took a long time to produce a skillful painter. This aspect of complexity can be measured in terms of a quantity called "depth." Depth can be defined in different ways,[16] but the essential idea is that it reflects the number of steps in the process leading from a system's origin to its current state of complexity.

"Our instincts for conservation in nature or preservation of historic monuments or any of those things are deeply connected with these notions of depth," said Gell-Mann. "Basically they can be summarized in the slogan: Protect Depth."

Complexity and Information

Depth, it seemed to me, was a way not only to measure complexity, but to measure information. A person, or a painting, both contain a record of all the information that went into making them. Human DNA stores a record of the evolutionary process producing the species; the human body records the developmental steps producing the individual. A painting is the sum of the artist's brushstrokes. There is complexity in a person and in a painting, and there is also information.

These thoughts revive the nagging realization that information as measured in bits and described by Shannon's information theory is not really the same thing as information the way people usually talk about it. As Warren Weaver wrote half a century ago in explicating Shannon's theory:

> The word information, in this theory, is used in a special sense that must not be confused with its ordinary usage. In particular, information must not be confused with meaning. . . . This word information in communication theory relates not so much to what you do say, as to what you could say. That is, information is a measure of one's freedom of choice when one selects a message. . . . The concept of information applies not to the individual messages (as the concept of meaning would) but rather to the situation as a whole.[17]

In other words, Shannon's information has to do with the probability of receiving a particular message, which depends on how many possibilities there were to begin with. It has nothing to do with whether the message would be meaningful to a complex adaptive system.

This all means that there's a difference between information quantity and information content. That is a critical point in understanding the nature of complexity and information processing by complex adaptive systems. There is more involved than just the information quantity that Shannon's bits would measure. The proverbial monkeys could type long enough to fill up a floppy disk with gibberish, containing maybe a phrase or two out of Shakespeare, but most of it unintelligible nonsense. You could fill up the same floppy disk with Shakespeare's plays, or an encyclopedia, or the NFL, NBA, NCAA, and major league baseball record books. The quantity of information would be the same, the content quite different.

This distinction between information quantity and content has a lot to do with the ability to compress the description of a complex system. If a long string of bits can be compressed into a short computer program, there is a lot of regularity, but not very much information. From that point of view, high regularity means low complexity, and low information.

But what about a message that can't be compressed? Such a message would be a random string of bits, for example, with no regularity. That message would take up the most space on your hard drive, so it would have lots of information, in Shannon's sense. But if it's a string of random digits, it wouldn't have much in the way of information content.

Pondering these questions over coffee usually gets me nowhere, especially since I don't drink coffee. So eventually I went to see someone who has pondered them much more profoundly, the physicist Jim Crutchfield. One of the key characters in James Gleick's famous book *Chaos*, Crutchfield had gone on to explore the foundations of complexity and computing, spending much of his time at the Santa Fe Institute, complexity theory's capital.[18]

He was at the University of California in Berkeley when I went to visit him in January 1995 and asked him to discuss the relationships connecting information, computation, and complexity.

"These are just words of course," he responded. "It really depends on who you talk to. The most important thing is that the words only take on meaning in a certain context and in certain uses. There's actually an interpretation of those words where they're all sort of the same thing."[19]

Well, that would certainly make everything a lot simpler, I thought to myself. I guess if you define complexity in terms of computing, and define information in terms of complexity, they would all turn out to be in essence the same thing.

Indeed, when Gell-Mann spoke of "ideal complexity," he was really referring to a quantity known as algorithmic complexity, or al-

gorithmic information content, a concept invented by Andrei Kolmogorov in the 1960s and developed further by Gregory Chaitin and
Ray Solomonoff.

Kolmogorov, Crutchfield explained, was really exploring the
foundations of probability theory in terms of Turing machines.
Looked at in the right way, Kolmogorov's complexity is just another
way of expressing what Shannon showed about information. Shannon, remember, defined the information produced by a source
in terms of probability of getting one message from among all the possibilities. Shannon's information measure is thus very much connected to randomness—randomness in a message means lots of
information.

"Kolmogorov . . . in a way deepened the foundations by saying
how a deterministic device [a Turing machine] could produce randomness," Crutchfield said. "But it turns out that his notion of complexity is really the same thing as Shannon's notion of information
produced by a source."[20]

So there you have it—information, complexity, and computing
with a Turing machine. "In the sense of understanding how random
and unpredictable a system could be, those three words could all sort
of apply equally well," Crutchfield said. And so everything was clear.
Until he added, "That's not my view."

Oh. I should have known. Pure randomness produces a high-
information quantity for Shannon, and pure randomness also
dominates measurements of algorithmic complexity (or algorithmic
information). To Crutchfield, complexity ought to reflect something
of an object's structure. The nonrandom stuff.

"Basically my current approach is there's something else that's
equally interesting and probably eventually will be more important
than just measuring degrees of randomness in a system, and that I describe with the word complexity," Crutchfield said. "Which is maybe
a bad choice because of this confusion I just mentioned, but here I
am. I use the word complexity—and a few other people do too—as a
moniker for the kind of structure an object has." [21]

Crutchfield has tried to capture his notion of structural complexity in a mathematical quantity he calls "statistical complexity." Statistical complexity is low when randomness is very high, but it's also low

when randomness is very low—that is, when the regularities can be condensed a lot. But in the intermediate range, where there's randomness mixed in with regularity, the statistical complexity is high.

"This structural complexity is low at either end of this randomness spectrum," Crutchfield explained. "As it gets larger it tells you that there is sort of more memory in the process."

Suppose, for example, that the systems in question produce sounds. One system might just be a repeating doorbell, sounding out ding-dong over and over again. That would be a very regular pattern, with low randomness, easy to describe simply. If, on the other hand, you had monkeys striking random keys on a piano, it would be very hard to find a pattern—randomness would be high, and describing the monkey music would take a lot of information.

From Crutchfield's point of view, these two situations are in a way quite similar. It wouldn't take long to figure out that ding-dong is just repeating itself. And it also wouldn't take long to figure out that somebody who knew nothing about music was striking piano keys at random. In both cases, the complexity would be minimal. So Crutchfield's statistical complexity quantity would be low in both cases.

But suppose you had a capable human piano player. It would take a long listening time to identify all the structure in a real piano piece. So the complexity would be high.

Technically, statistical complexity measures how much memory you need to record enough information about the system to predict what it will do next. "You've got this question, how long does it take me, as I begin to make the measurement, to come to that knowledge that I can predict?" Crutchfield said. "If it's a pretty random source, very quickly I realize I can't predict it. So it's highly unpredictable. . . , but there's not much I have to remember and not many measurements I need until I come to realize that. So it has low statistical complexity."[22] If it's a completely regular source, like the ding-dong doorbell, it is highly predictable after a short time, requiring very few measurements. Therefore that statistical complexity is also low.

"In between there are these more complex processes; they're more structured," said Crutchfield. "You have to remember more things" to be able to predict the system's behavior. So structural complexity in those situations is higher.

It's important to note that when measured in this way, complexity is in the eye of the beholder, or the observer or IGUS. (Crutchfield calls such an observer an "agent.") Statistical complexity measures an agent's ability to predict its environment. Thus statistical complexity measures something that is meaningful for the agent. "So part of this actually is moving in the direction of trying to understand what information means," Crutchfield acknowledged.

Now this is a rather bold step. Crutchfield is trying to do what Shannon did not do—describe information content, not merely information quantity.

"Part of the colloquial sense of the word *information* is, and probably the more important aspect of it is, the notion that information means something," Crutchfield commented. "Information theory as written down by Shannon is not a theory of information content. It's a quantitative theory of information. All it tells you is how much information is there, not what the information is."[23] Statistical complexity, on the other hand, tries to capture what information means to an observer, or agent.

Crutchfield's agents are, of course, complex adaptive systems. They survive by making observations of their environment and representing regularities in that environment in their memory—in other words, they make what Gell-Mann calls a schema.

Here's where it gets complicated. Because no agent is an island. When an agent looks out into its environment (to make a schema describing it), what that agent typically sees is a bunch of other agents.

Think of it as a jungle out there. Surviving in the jungle depends on being able to figure out what all the other animals are likely to do. (It also helps, of course, to recognize other regularities in the environment, such as when it will get dark and where water always flows.)

Now, any agent's (or animal's) ability to make predictions about life in the jungle depends in part on how smart that animal is—or in other words, the schema-making resources it possesses. An agent with a vast memory storage system can record many more nuances about the jungle, identify subtle regularities as well as obvious ones, and use that extra information to make better predictions—such as where food is likely to be found. Further useful resources would include a fine-tuned sensory apparatus for acquiring information—say, sharp eyes and a keen sense of smell. The more resources available, the better one agent will be able to figure out the other agents.

As Crutchfield notes, though, the precise nature of the resources isn't as important as what the resources can do—computationally. An agent's success depends on how well it records and processes information about the environment in order to choose wise behaviors. And the law of the jungle says that whoever chooses the best behaviors survives. The losers go extinct.

Such survival competition is, of course, what biological evolution is all about. So Crutchfield sees in his view of agents and computing resources a new picture of evolution—what he calls "evolutionary mechanics." In other words, he thinks his ideas can explain how new species emerge. It's simply a matter of looking at how all those animals in the jungle process information in order to compete with one another for survival.

Suppose you're the agent, and one agent you might have to interact with is a Jeep. (Remember, you're just an animal in the jungle, so as far as you know the Jeep is just another animal.) You want to be able to predict the Jeep's actions (that is, make a schema or model that contains information about regularities in the Jeep's behavior). The first step in making a schema about another agent is figuring out that it can exist in various conditions, or states. After observing a Jeep for a while, you might decide it could exist in two states—"on" (moving) or "off" (parked). You would then have a two-state model of Jeeps. But if you observe a little longer, you will eventually discover that the on-state (moving) has different substates, such as moving forward or backward. If you are a really sophisticated agent, with great powers of observation and memory, you might infer several different states of forward motion as well (different gears, four-wheel drive, etc.). A less sophisticated agent, however, may not be able to detect or model such subtleties, and might be more likely to get run over.

Crutchfield explored these ideas in an information-rich paper called "The Calculi of Emergence," a deep, difficult but extremely insightful paper, spelling out in a rigorous way the principles and the math that can describe the world in computational, information-processing terms.[24] In his view the universe is a deterministic dynamical system, its parts all moving according to natural law and the forces that act on them. Agents within this universe do not appear to be deterministic, though, to any agent observing them. That's because the observing agent has limited resources to make a model or schema of other agents. Since one agent cannot predict everything about another agent, some of that other agent's actions seem random. With more computational resources, an agent could predict more about another agent's actions and therefore choose wiser behaviors in response. An agent's model of the jungle, after all, controls how the agent will behave in response to various sensory inputs.[25]

Viewing living agents in this way, you can see how new features of life could emerge, Crutchfield says. Specifically, agents can adapt to their environment—and improve their ability to survive—by changing their architecture for processing information. Or in other words, improving their ability to compute.

With more resources, an agent can store more information about the environment and make more sophisticated models. So it would seem that the best strategy for survival would be to amass resources. But resources are not always available. An agent that can get by with a simpler model, requiring few resources, will have a competitive advantage in times of resource scarcity. So there is a tradeoff between the need for more resources and the need to make a more sophisticated model of the environment.

That's why some simple agents can find their own niche in a complex world. A niche is basically "a subset of environmental regularities" in Crutchfield's terms. In other words, it's a smaller, simpler world within the larger complex world, a patch of the jungle where the demands for predictive power are small. By discovering a simple model that describes that niche well, some agents can survive quite nicely. This tradeoff between complex models and the need for resources explains why some creatures with limited computational resources, like a fly, can coexist in a universe with an organism possessing vastly greater computational resources, like Bill Gates.

The question is, how do these different species develop? Not from a biological standpoint, but from an information-processing standpoint? The answer, says Crutchfield, is through "innovation," in which an agent learns to use its limited resources to make a new, more efficient model. An agent that makes a new, improved model of its environment then behaves differently. It becomes something new.

At the core of Crutchfield's views is his realization—or perhaps his appreciation of what earlier thinkers have realized—that reality is a process. It's a process balancing stability and order against instability and change. At the end of his paper, Crutchfield quotes Alfred North Whitehead, the mathematician-philosopher-theologian: "The art of progress is to preserve order amid change, and to preserve change amid order." In Crutchfield's universe of computational agents, information storage assures stability and order; information processing permits instability and change. Information storage and processing—that is, computation—is therefore the paradigm of reality as a process.

There's an additional point that Crutchfield does not emphasize—but that struck me as interesting: Regularity in the environment does not seem to be something just out there to be discovered. The regularity observed depends on the resources of the observer. In other words the description of the universe depends on its observers' ability to describe the universe.

In the case of people, that description is called science. And since the human brain is not an infinite resource, it can't really create an omniscient description. When a new computational paradigm comes along that makes better use of available resources, an agent builds a better model of its reality. In the same way, scientific revolutions occasionally occur and change the human view of the nature of the universe. This process should continue as brains add resources, like new observing instruments, which is no doubt one of the reasons that new technologies drive advances in science. They provide new data, and new conceptual resources for making a more efficient model of the data.[26]

I posed the point to Crutchfield:

ME: "There's a tradeoff between the amount of resources you need and amount of complexity you can detect. It seems to me you've got situations where the agent determines a lot about what kind of regularity or structure gets noticed in its environment. . . . So it's not something that's strictly a question of the agent determining properties of the environment, it's almost the way the agent is built imposes the regularities that it sees, right?"

CRUTCHFIELD: "Yeah, that's true. I think so far I haven't been called to task for this. Because I think there's a way to misinterpret some of the deductions you would make from that. The bad interpretations would be, 'Oh, he's talking about wanton subjectivity, everything is relative, cultural relativity.' That's actually not it. What I'm interested in is actually an objective theory of subjectivity. . . . The problem is, which I think this development illustrates, is that what I can understand about this system depends on me."[27]

From this point of view, I think it's fair to say, the observer does assume a certain importance in the description of reality. But where does that leave the role of the observer in quantum physics—in existence itself—that inspired all this concern with observers to begin with? For that we must go back to Murray Gell-Mann.

"Let's talk about quantum mechanics," he said that afternoon in Chicago. "In quantum mechanics there's been a huge amount of mystical nonsense written about the role of the observer. Nevertheless, there is some role for the observer in quantum mechanics, because quantum mechanics gives probabilities. And you want to talk about something that is betting on those probabilities. That's the real role of the observer."[28]

In other words, life is a casino. To win, you have to understand quantum mechanics.

QUANTIFYING INFORMATION

Shannon entropy
   (or Shannon information)
   Negative logarithm of probability of a message. A measure of uncertainty or freedom of choice in composing a message.

Algorithmic information content
   (or algorithmic complexity, Kolmogorov complexity)
   The number of bits in the smallest program that outputs the message string when run on a universal Turing machine. Dominated by randomness.

Logical depth
   The number of steps in a deductive or causal path connecting a thing with its plausible origin. Or the time required by a

universal computer to compute the object in question from a program that could not itself have been computed from a more concise program. A measure of organization.

Statistical Complexity

The amount of memory in bits required for a machine or agent to predict its environment at a given level of accuracy. Or the minimum amount of historical information required to make optimal forecasts of bits in an object at a specified error rate. A measure of structure.

Chapter 9

Quantum Reality

The discovery of quantum mechanics is one of the greatest achievements of the human race, but it is also one of the most difficult for the human mind to grasp. . . . It violates our intuition—or rather our intuition has been built up in a way that ignores quantum-mechanical behavior.

—MURRAY GELL-MANN,
The Quark and the Jaguar

The only "failure" of quantum theory is its inability to provide a natural framework that can accommodate our prejudices about the workings of the universe.

—WOJCIECH ZUREK,
"Decoherence and the Transition from Quantum to Classical"

If you don't think it's important to know something about quantum mechanics, just listen to Murray Gell-Mann.

"It requires a certain degree of sophistication . . . to grasp the existence of quantum mechanics," he says. "I would say there's much more difference, from this point of view, between a human being who knows quantum mechanics and one that doesn't, than between one

177

that doesn't and the other great apes. The big divide is between people who know quantum mechanics and people who don't. The ones who don't, for this purpose, are goldfish."[1]

Don't worry. Unless you skipped chapter 1, you've already advanced well beyond the goldfish stage of quantum understanding, perhaps to the dolphin level. Just remember, quantum mechanics is like money, it's like water, and it's like television. (That is, energy comes in units, like pennies; particles can be waves, like ice cubes or water; and reality has multiple possibilities, like different TV channels.)

Of course, deep down you no doubt knew that it couldn't really be that easy. And it's not. I left out an important, deeper, more complicated aspect of quantum mechanics. In addition to being like money, water, and television, quantum mechanics is like the movie version of the board game Clue.

Avid film fans know that *Clue* (Tim Curry, Martin Mull, Paramount Pictures, 1985) was unusual—it was not really one movie, but three. Different theaters showed different endings. Each ending had to be consistent with what had happened previously in the film, of course. But when it became time to reveal who killed the victim, Mr. Boddy, various possibilities remained. In one version, Mrs. Peacock did it; in another, it was Yvette, the maid; and in the third it was Mr. Green, in the hall, with the revolver.

And that's very much the way that quantum mechanics works—multiple possibilities, waiting to become real. In any one theater, only one of the suspects turned out to be Mr. Boddy's murderer. In real life, only one of the many possible quantum realities comes true. The question is, how.

In chapter 1, I merely noted that an observation plucks one reality from among the many possibilities. But that can't be the whole story. When you leave a room, the chairs don't rearrange themselves, even though a pure quantum description would allow each of them to occupy many possible locations, just as electrons do. But chairs and other big objects stay put. "This is to me the main problem with quantum mechanics," says Wojciech Zurek, the Los Alamos theoretician. "In the real universe, we don't see chairs being spread all over the place."[2]

Quantum mechanics does not explain why. And as you may have noticed, so far, neither have I. But you can hardly blame me. For seven decades experts and amateurs alike have argued about this issue. Resolving it requires something more than just the quantum math—it requires an interpretation of the quantum math. Many such interpretations have been proposed over the years, all controversial. (Sometimes it seems like there are as many interpretations of quantum mechanics as there are different possible quantum realities.)

Basically, the interpretation problem boils down to a simple paradox: Quantum math describes a world very unlike the world we live in, yet it gives all the right answers to any experiment we can do. We live in a "classical" world, with one reality, yet the quantum math contains many possibilities that we never encounter—multiple realities, like all those different TV channels. But we see only one show at a time. There is no picture-in-a-picture.

Niels Bohr, the godfather of quantum theory in its early days, deferred the mystery to later generations by devising a framework for doing quantum physics without worrying about that stuff. Forget about the "quantum world," Bohr said. We live in a classical world, and we can communicate to each other only in classical terms. We do experiments with classical devices. We can cope with quantum mysteries if we remember that we must always describe our results in common language and include in our description the experimental apparatus we use. Along that path there will be no contradictory experiences. True, an electron can behave like a wave or like a particle, but never both at the same time in any specific experiment.

Bohr's philosophy, built on what he called the principle of complementarity,* served physics well for a long time. It made progress in quantum physics possible. I think it was similar, in a way, to Newton's introduction of absolute time and space as a backdrop for his laws of motion and gravity. Newton no doubt knew that space and time were not, in reality, so simple. Yet stopping to understand them fully first would have retarded further progress in the rest of physics.[3] Bohr did much the same thing, providing a way to make quantum mechanics useful while avoiding some embarrassing questions.

---

*The essence of complementarity is that a complete description of nature requires the use of mutually exclusive, yet complementary, aspects, such as wave and particle.

Even so, Bohr's complementarity principle, the keystone of what came to be known as the Copenhagen interpretation of quantum theory, never satisfied everybody. Einstein was the most conscientious of complementarity objectors; he waged a three-decade debate with Bohr that ended only with Einstein's death in 1955. Others also tried to find better ways to explain the quantum mysteries, too many ways to recapitulate here. But one in particular stands out as a testament to the depth of the problem—the proposition that all those multiple quantum possibilities really do exist after all. In *Clue*, three possible endings all came true, in different theaters. Maybe in real life, all the quantum possibilities come true too—in different universes.

Many Worlds

This sounds like a theory straight out of science fiction, a tale of countless parallel universes to accommodate all the possibilities that quantum physics contains. Actually, it's straight out of a Princeton doctoral thesis written in 1957 by Hugh Everett III.[4] In the words of physics historian Max Jammer, it is "undoubtedly one of the most daring and most ambitious theories ever constructed in the history of science."[5]

Everett's theory is known as the "many worlds" or "many universes" interpretation of quantum mechanics. The standard illustration of how it works invokes a famous paradox, involving a cat, proposed by Erwin Schrödinger in 1935. Among scientists Schrödinger's Cat has become more famous than Morris or Garfield. But I prefer to describe the situation with my sister's cat, Fibby.

Fibby is all black and allows visitors to pet her only three times before she bites them. Consequently Fibby is an excellent candidate to lock in a box with a vial of cyanide, as Schrödinger envisioned. The cyanide vial is connected to an elaborate apparatus containing a small lump of radioactive material, a Geiger counter, and a hammer. (Schrödinger also points out that this apparatus must be "secured against direct interference by the cat.")[6]

The amount of radioactive material needs to be very small, so that over the course of an hour there is only a 50-50 chance that even one atom will decay. If it does, the Geiger counter will click, activating a switch that releases the hammer, which smashes the vial, poisoning Fibby.

Once Fibby is locked in the box, though, nobody knows when the atom will decay. So the question is, after an hour, is Fibby dead or alive? The answer is you don't know unless you open the box and look inside. This is a troubling answer, more so for physicists even than for cat lovers, because physicists like to think that they can predict the future. Yet the quantum equations describing radioactive decay can't predict what you'll see when you open the box, just the odds that Fibby is dead or alive. The quantum description of the situation suggests that Fibby is in limbo, half dead and half alive, until somebody opens the box. Schrödinger argued that quantum mechanics is therefore faulty, because it makes no sense to say that a cat is alive and dead at the same time.

But if parallel universes exist, the situation changes. Sure, you might open the box and find Fibby alive. But in some other universe Fibby might be a goner. What you see in the box depends on which universe you happen to find yourself in at the time. Put another way, whenever an observation is made, new universes are created, corresponding to one from among all the possible outcomes.

That, at least, is the way the Everett theory is generally summarized. But it's a crude characterization. Not all quantum physicists would express Everett's idea in this way. Some would say it's really not so much a matter of creating new universes; it's more like the multiple universes are all there to begin with, contained in the equations of quantum mechanics.

Everett himself, in fact, did not speak of "many universes" in his thesis or in a longer account published in 1973. Actually, he spoke of one universe, described by the standard equation of quantum mechanics (known as the Schrödinger equation). But within that universe, certain subsystems (like observers) interact with other subsystems, forging connections (or correlations) between the observer and other parts of the universe. The trouble is, the Schrödinger equation contains those pesky multiple possible realities. So one observer can be related to others in many different ways. Anytime two parts of the universe interact, Everett concluded, the different possible results of that interaction all become real.

Physicists use the term *superposition* to describe this situation of multiple possibilities existing simultaneously. A cat can be in a super-position of dead-and-alive; a chair can be in a superposition of loca-tions, spread out over a room; or a coin can be in a superposition of heads-and-tails (the basic idea of a qubit in quantum information theory). In Everett's theory an observer's interaction with other sys-tems puts the observer into a superposition of different relationships. (See the discussion of Everett's theory at the end of this chapter.)

Seen in this way, Einstein's worry about an observer (such as a mouse) changing reality takes a different form, Everett said.[7] The mouse doesn't change reality by looking at it, but merely participates with the rest of reality in different ways—simultaneously. When the mouse looks at the universe, the universe carries on. But the mouse changes—splitting into a new superposition of states—after inter-acting with the rest of the universe. "The mouse does not affect the universe," wrote Everett. "Only the mouse is affected."[8] In a sense, new copies of the mouse are created every time it observes some-thing. In a similar way the cat in the box splits into two copies—one purrs happily and the other dies of cyanide poisoning. Every time a cat, or a mouse, or any other observer interacts with the rest of the universe, yet another copy of that observer pops into being to reflect the new possible realities contained in the quantum equations. In other words, observations don't create multiple new universes, but multiple new observers—within the one universe we all occupy. To the objection that no observer "feels" this splitting into multiple copies, Everett responded that nobody notices the rotation of the Earth, either.

The catch to all this is that the new copies of any observer remain utterly unaware of, and unable to communicate with, the other copies. For all practical purposes, the new copies are off in worlds of their own. That is why Everett's followers have generally described the universe as splitting into new realms. Eventually every conceiv-able result of every possible act of observation becomes real in some realm or another. It's easy enough to think of those realms as multi-ple universes.

Needless to say, the many-universes theory is not universally ac-cepted. After all, it's hard to believe that in some other universe the Buffalo Bills have won four Super Bowls. Even John Wheeler, who supervised Everett's thesis, no longer supports the many-universes idea, because "its infinitely many unobservable worlds make a heavy load of metaphysical baggage."[9]

Still, many physicists took Everett's idea seriously. Bryce DeWitt of the University of Texas championed it in a 1970 *Physics Today* article.[10] Later David Deutsch of Oxford University in England became a leading advocate of the many-worlds view, which he extolled at length in his 1997 book *The Fabric of Reality*.[11] At a 1997 conference on quantum mechanics, an informal poll found 8 supporters for the many-worlds interpretation out of 48 in attendance. It came in second to the Copenhagen interpretation, with 13 votes. Actually "undecided" really came in first, with 18 votes, an indication of how up in the air the interpretation of quantum mechanics still is.[12]

Nevertheless, the many-worlds view has its drawbacks. Like the anthropic principle, it is of limited usefulness. And it really doesn't eliminate the central mystery—if all these universes exist, why is it that we ordinarily notice only one of them? How does the foggy world of atoms become the rock-solid reality of ordinary life? In other words, what is the connection between quantum physics and classical physics?

Oddly enough, nowadays many physicists believe that this question has an answer. Rooted in Everett's views, new approaches devised in the 1980s and developed further in the 1990s have produced a way to connect quantum physics and common sense. At the heart of this new understanding is a process called quantum decoherence. From the decoherence viewpoint, the various possible realities described by quantum theory all really do exist, but only very briefly. Only one reality lasts long enough to be noticed. The rest die, like unfit species that go extinct.

The secret to this "selection" of realities is information processing, of course. Information about a system is recorded by the environment, and in that process one reality emerges. The principle of selection that chooses that reality is not fitness, as in natural selection of species during evolution, but consistency. The reality that becomes concrete has to be consistent with all the previous realities selected from the quantum fog; that is, reality must be consistent with all the information about the past that is recorded in the environment. This "information" is stored in the environment in the form of the movements of atoms or light waves or other particles that encounter the object in question. *Decoherence* is the technical name for this disappearance of the multiple additional possibilities. Ordinary existence as we perceive it, therefore, arises from quantum decoherence.

For example, quantum math applied to a chair describes it as a fuzzy state spread over different rooms in the house. Nevertheless, anyone looking for a place to sit would always find the chair in a single place. And the reason we do is because the interaction of a chair with its environment creates a record of the chair's position. In a sense, the environment "selects" a single position from among the possibilities.

When I had lunch with Wojciech Zurek in England in 1990, he outlined the basics behind the decoherence process, and the next year he published a sort of decoherence manifesto in *Physics Today*. By 1993 I'd decided to write an article about decoherence, so I went to Los Alamos to ask him to explain it to me in more detail.

He likes to call his approach the "existential interpretation" of quantum mechanics (a name that as far as I can tell hasn't really caught on). "It's sort of an attempt to build a bridge between the Everett many-worlds interpretation on one hand and Bohr's Copenhagen interpretation on the other hand," he said. "And the chief pillar which supports that bridge is the decoherence process."[13]

Some of quantum theory's early pioneers understood some aspects of decoherence, but the modern enthusiasm for this approach accelerated only after a 1985 paper by Erich Joos and H. Dieter Zeh showing that a single distinct location of an object emerges naturally from interaction with the environment. Molecules or light particles bouncing off an object carry away information about that object's shape and position. Just as a pebble dropped in a pool sends out waves recording the pebble's presence, our eyes see light that has bounced off various objects, revealing their size, shape, color, and location. During the early 1990s, Zurek and others published a series of reports further analyzing the mathematics of this process, showing that monitoring by the environment singles out only one of the quantum possibilities, or states, to be accessible to our senses.

"There's nothing mysterious about it," Zurek assured me. "All of the normal objects that we think of as classical, chairs, tables, bacteria, even things which are fairly small, leave an imprint on the envi-

ronment. The interaction of the object with the environment destroys certain states much more rapidly than others." The possibility
that survives is always the most familiar sort, such as a solid chair real
enough to sit in.

In fact, it's possible to calculate just how long it takes for the
many possibilities to freeze into one. The process depends on the
mass of the object, the temperature, and distances involved. At very
small distances, very low temperatures, and very tiny masses, decoherence can take a while, which is why electrons in atoms seem to behave so strangely. But for objects the mass of a chair, at ordinary
temperatures, the decoherence takes place faster than it takes light
to cross an atom. While different possibilities for a chair's location
may exist, they disappear much too rapidly for anyone to ever notice.
The same principle applies to any big object. (Unless, of course, that
big object is somehow isolated from the environment and kept at a
very low temperature.)

Consider, for example, the position of the moon. A quantum description suggests that the moon, like an electron, could be in many
places at once. But light from the sun interacts with the moon.
Roland Omnès of the University of Paris has calculated that the sunlight interaction freezes the moon's specific location from among the
various possibilities in ten-trillionths of a trillionth of a trillionth of a
second. So everybody sees the moon in the same place. [14]

Beyond explaining human perception of reality, the decoherence
process may also offer a new way to understand the direction of time.
In quantum mechanics, as in classical physics, time is free to flow forward or backward. In real life, though, time always seems to move forward. The world is full of irreversible actions. Popcorn never unpops,
and eggs never unscramble. Physicists do not all agree about how to
explain why this "arrow of time" always points in one direction.

It may be that the irreversibility of events in the world is related
to quantum decoherence, since it seems that the transfer of information from a system to its environment would be impossible to reverse.
Omnès has pointed out, however, that quantum mathematics does in
principle permit reversing the time direction. But it would be an immense task to prepare the proper conditions for making events run
backward. You would need an apparatus so large that the universe
could not contain it. So in fact, reversing the flow of quantum

decoherence would be impossible in practice. Nevertheless, it's fair to say that the time-direction question needs more study. As do other issues about quantum physics and reality.

Consistency

Decoherence itself is not the whole story of quantum reality. It merges with another approach to the problem, pioneered by Robert Griffiths of Carnegie Mellon University in Pittsburgh, called "consistent histories." Griffiths' idea, introduced in 1984 and refined in the 1990s, conceives of the history of a quantum system as a sequence of properties, like a series of snapshots that could be compiled to make a movie. While a system might start out with many possible arrangements, after a while only a few series of snapshots would make any logical sense. Only "consistent histories" are meaningful—like in the movie version of Clue, where all of the endings of the story had to be consistent with the earlier parts of the film.

The consistent histories approach has been extended and embellished in detail by Murray Gell-Mann and his collaborator Jim Hartle, who showed how to connect the ideas of decoherence and consistency. They regard their approach as "an attempt at extension, clarification and completion of the Everett interpretation."[15]

They point out that from among all the quantum possibilities, different sets of consistent histories can emerge in the universe. Humans experience one reality, and not many, because they participate in only one line of history. This view seems to fit in nicely with Zurek's picture of decoherence—as Zurek says, the decoherence process assures that you will see consistent histories. But Gell-Mann insists that the situation is much more complicated. After his talk in Chicago, I approached him and asked about the difference between his and Hartle's approach and the approach of Zurek and friends. "The difference is that we do it right and they do it wrong," he replied. I could tell that I would have to discuss this with Gell-Mann at greater length sometime.

So a year or so later I flew to Chicago to have dinner with him (it was the only place where he would have time to talk to me, he said), and we explored the issue in detail. In his view, decoherence arises naturally when the quantum description of the universe is analyzed on an appropriate scale (the "coarse-graining"). There is no need to divide the universe into subsystems interacting with their environment, as Zurek does.

"We talk about the whole universe, and if there are any observers they're inside," said Gell-Mann. "He [Zurek] talks as if there were such a thing as an environment. . . . He talks as if God has ordained that certain things are followed and other things are environment."[16] But the universe doesn't have an environment. Applying quantum mechanics to the whole universe should produce a way for classical realities to emerge inside it without any mention of an environment, and without any special role for an observer, Gell-Mann insisted—a point he made during his talk in Chicago in 1992.

"Many of us over the last thirty-five years have been developing the modern interpretation of quantum mechanics in which the role of the observer is not so great," he said then. This "modern interpretation" did away with the mysticism of the past and applied quantum mechanics to the whole universe. There was no restriction to describing laboratory experiments, as in Bohr's Copenhagen interpretation.

"It's not wrong, but the Copenhagen interpretation is extremely special," Gell-Mann said. "But nowadays for quantum cosmology we want a quantum mechanics that applies generally, not just a physicist repeating a physics experiment over and over again in the laboratory. The Copenhagen interpretation carried with it a lot of baggage having to do with the importance of the observer, and people claimed it had to be a human being, it couldn't be a goldfish, all sorts of nonsense."[17]

Nevertheless there are observers around, and they do have a part to play in the quantum drama. Quantum mechanics provides the world with probabilities for various futures. "Somebody's got to use those probabilities," Gell-Mann pointed out. That is, somebody has to bet on them. And that is exactly what IGUSes do. They are complex adaptive systems that act as observers, making bets about how best to survive.

To Gell-Mann, the IGUS-observers can exist precisely because there are "probability tracks" within the universe that at certain times and places offer conditions favorable for the evolution of complex adaptive systems. Those regions have enough regularities to permit the existence of something that can exploit those regularities. Regions that support the existence of observers are realms in which the probabilities offer many safe bets—almost certain bets—so a complex adaptive system could make a useful schema for describing how its environment behaves. Those "almost certain" probabilities are the hallmark of classical Newtonian physics. Thus the regions of spacetime with quantum probability tracks offering such good bets are called "quasiclassical realms."[18] The features of a quasiclassical realm result from the natural play of the laws of physics as the universe evolves over time from its starting point.

"There are huge regions of spacetime and many probability tracks where there aren't any observers," Gell-Mann noted. "For example early in the universe there certainly weren't any. But gradually as quasiclassical variables emerge from the quantum fog, at least along certain of the probability tracks, you begin to get enough regularity so that something could evolve that exploits this regularity in its environment."[19]

Even within a quasiclassical realm, many quantum histories are possible, Gell-Mann points out. But we only experience one of them. He compares the situation to an afternoon at the races. With eight races, and 10 horses per race, 100 million scenarios are possible for the winners list, with the odds of some outcomes better than others. But as the races are run, definite results are recorded and other possibilities disappear. "In the experience of any one thing, only one of those histories is experienced," he said. "The others are just completely not there. Once the horse race is over, one horse has won."[20] All the possibilities of other horses winning are no longer meaningful.

Before you start thinking that quantum mechanics is as simple as a day at the races, it's important to realize that a lot goes on at the races that nobody pays any attention to. Stray atoms here and there floating inside a horse's nostril are unlikely to affect the outcome of a race. Many subtle variations in what happens go unnoticed. So a day at the races is not really a single history, but a whole set of quantum histories that end up with the same list of winners at the end of the day. It's not just one history, it's a set of "consistent" histories. In the same way we all live within not one history, but within one set of consistent histories. In other words, our quasiclassical realm is a set of histories so similar that we ordinarily don't notice the different fuzzy quantum possibilities. Therefore we think we live in a classical universe.

The set of histories within a quasiclassical realm are all consistent with one another at the level of coarse graining used by systems that gather and use information (whether people, or computers, or the crystalline entities of *Star Trek* episodes). Life evolves and adapts to the quasiclassical realm, not the underlying fuzzy quantum reality. That is why people can catch baseballs and birds can navigate by the positions of the stars. In other words, life inhabits a quasiclassical realm because that's the only kind of realm in which predictions are possible, and if predictions aren't possible, complex (living) systems could not adapt and evolve.

The full quantum description of the universe contains many different sets of histories—in other words, many realms. But not all those other realms would be quasiclassical like ours. In fact, Gell-Mann says, most other sets of histories would probably not include the regularities and patterns that make science—or life—possible. So perhaps our quasiclassical realm is unique, and any forms of life (or other type of complex system capable of recording and using information) would have to live in the same realm that we do.

Then again, maybe not.

Gell-Mann and Hartle once conjectured that other sets of quantum histories might permit lifelike systems to emerge in quasiclassical realms other than our own. "In science fiction," Gell-Mann said, "those are called goblin worlds." Communication between these systems and our own quasiclassical realm seems to be, in principle, possible. If so, the IGUSes from a goblin world might figure out that we exist as well and try to send us some quantum e-mail.

"If two essentially distinct quasiclassical realms exist, they may overlap, in the sense of having some features in common," Gell-Mann and Hartle wrote in their paper exploring this possibility.[21] "It is then possible that IGUSes in different but overlapping realms could make use of some of the common features and thus communicate." In principle, listening for signals from cousin IGUSes would not be very different from searching for signs of extraterrestrial life. We would expect to see some coded signals spelling out the digits of pi—or as in the film *Contact,* a sequence of prime numbers. Such messages might emerge in an elaborate quantum mechanical laboratory experiment, in which cousin IGUSes from another realm influenced the quantum properties being measured in our lab. Of course, our cousin IGUSes could be sending us other kinds of messages that we don't know how to recognize.[22]

Or the whole idea could be misguided. Not all quantum physicists share Gell-Mann and Hartle's view on the quantum description of the universe. And even some who do find the idea of communication with other quasiclassical realms to be quasicrazy. But in quantum mechanics, crazy things are often correct. "If the universe exhibits IGUSes in realms essentially different from the usual quasiclassical one, that does not constitute a paradox," wrote Gell-Mann and Hartle, "but rather an intriguing example of the richness of possibilities that may be shown by a quantum universe."[23]

Complete and Consistent

Goblin worlds aside, research into decoherence and consistent histories has put a new face on the problems of quantum mechanics, and many experts see a smile on that face. Omnès, for example, has published extensively on the new interpretation of quantum mechanics that has emerged from all this work.[24] He summed up the situation in 1992 by declaring the search for an interpretation of quantum theory to be over.

"There is now a workable complete and consistent interpretation of quantum mechanics," he wrote.[25]

Omnès himself has extended Griffiths' consistent histories approach to analyze the logic behind quantum physics and to demonstrate how the standard laws of cause and effect in the real world arise from quantum mechanical principles. But Omnès departs from some of the others in this field by dismissing the fundamental importance of information. The notion that "the only content of physics is information" is an "extreme positivistic point of view," Omnès contends. "Information theory is a very useful superstructure of science, but it cannot be its foundation."[26]

Zurek, however, is not impressed by Omnès's negativity about the importance of information. "I think he's dead wrong," Zurek told me. When I visited him at Los Alamos he elaborated on this point at length.

"In a sense, one could say that the only way we perceive the universe is by consulting what we know about the universe in our own memory. So that's obviously one place where this whole business of information, and one would probably even want to say information processing or computation, is crucial," Zurek said. And it is just this information-centered approach that makes it possible to deal with the existence of observers while avoiding the uneasiness of resorting to an anthropic principle.

"I think there is a way now to represent within physics systems which have all the basic features of us, the observers," Zurek said. "But one can do it without being unduly anthropic. In other words, instead of talking about a human being or consciousness or something or other, one can talk about information acquisition, information processing, information utilization."[27] It's the new appreciation of the importance of information, Zurek believes, that is making progress possible toward understanding deep issues about existence and reality.

"A lot of these questions, they tend to go back at some level to information," he said. "You cannot escape it. On one hand it's very annoying to a physicist, at least of the old school, because information seems to be such an elusive and intangible, and unphysical, almost, concept. And on the other hand I think it's pointing toward something that we should listen to. This information thing is really important."[28]

As I was finishing this chapter, a new paper appeared on the Internet emphasizing the role of information in quantum decoherence. J. J. Halliwell of Imperial College in London has contributed many papers to the quantum decoherence literature, and in this one he directly addressed the information question. He conjectured that the idea of decoherence occurring as the environment recorded information ought to be precise, not a mere hand-waving analogy. In other words, if information stored in the environment is responsible for the reality we see, the number of bits stored (by decoherence) ought to be related to the number of bits needed to describe reality (the consistent histories). Halliwell produced some calculations to indicate that, at least in certain specific cases, the relationship was in fact quantitative. "We thus find verification of our conjecture: the number of bits required to describe a decoherent set of histories is equal to the number of bits of information about the system stored in the environment."[29]

This work reiterates a very important point: the idea of reality being related to information is not a mere metaphor. Information is real. Information is physical. And understanding that is opening a whole new avenue for understanding the nature of existence.

Halliwell notes that his work in this area was inspired by Zurek's observations that ideas from information theory have not been exploited as much as they might be in physics. As a further source of inspiration, Halliwell mentions John Wheeler's idea of "It from Bit." It's only fitting, then, that the pursuit of information's importance in science should take us back to Wheeler and in particular to his favorite astrophysical character, the black hole.

Hugh Everett's Many Worlds Theory

Hugh Everett's "many-worlds" interpretation of quantum mechanics has itself been interpreted in many different ways, so it might be worth going into a little more detail about what he actually said. Everett did not coin the phrase "many worlds"; rather he called his approach the "relative state" formulation of quantum mechanics. His essential idea was to apply the Schrödinger equation to an "isolated system."

No system is truly isolated, except the universe as a whole. The universe can, however, be broken down into subsystems—in other words, the universe can be considered as a "composite" system. The physicist's task is to describe the universe's subsystems without resorting to some "observer" outside the universe to determine what happens.

A key point is that the universe's subsystems inevitably interact. Therefore no subsystem of the universe is independent of all the rest. In fact, Everett points out, any observation or measurement is just a special case of an interaction between subsystems. So it is impossible to describe a given subsystem completely—in physics lingo, describe its "state"—without regard to the rest of the universe.

Any subsystem can exist in many possible states, then, depending on the state of the rest of the universe's subsystems. "It is meaningless to ask the absolute state of a subsystem," Everett asserted. "One can only ask the state relative to a given state of the remainder of the subsystems."[30]

So far, it's all pretty straightforward. But the Schrödinger equation complicates things by allowing more than one pair of relationships between subsystems. In fact, the Schrödinger equation allows multiple possible combinations of subsystem states—what physicists call a "superposition" of states. If you take the Schrödinger equation seriously, Everett decided, you have to conclude that all the different superpositions of states are equally real.

"The whole issue of the transition from 'possible' to 'actual' is taken care of in the theory in a very simple way—there is no such transition," Everett commented. "From the viewpoint of the theory all elements of a superposition (all 'branches') are 'actual,' none any more 'real' than the rest."

In this approach, "observers" are just certain subsystems within the universe, systems that "can be conceived as automatically functioning machines (servomechanisms) possessing recording devices (memory) and which are capable of responding to their environment" (in Gell-Mann's words, IGUSes). As an "observer" interacts with other subsystems (or "object-systems," in Everett's terminology), their states become connected, or "correlated."

"From the standpoint of our theory, it is not so much the system which is affected by an observation as the observer, who becomes correlated to the system," Everett wrote.

"Correlations between systems arise from interaction of the systems, and from our point of view all measurement and observation processes are to be regarded simply as interactions between observer and object-system which produce strong correlations."

So any one "observer's" state will change after interaction with an object-system. But because there is more than one way to describe the observer-object-system relationship (a superposition of different ways), there is no single state that can uniquely describe the observer after the interaction.

"It is then an inescapable consequence that after the interaction has taken place there will not, generally, exist a single observer state," Everett commented. "There will, however, be a superposition of the composite system states, each element of which contains a definite observer state and a definite relative object-system state. . . . [E]ach element of the resulting superposition describes an observer who perceived a definite and generally different result." The correlations set up by multiple interactions ensure consistency among multiple communicating observers, Everett points out. But any one observer, after an interaction, exists in multiple states.

"Whereas before the observation we had a single observer state," he explains, "afterwards there were a number of different states for the observer, all occurring in a superposition. Each of these separate states is a state for an observer, so that we can speak of the different observers described by the different states. On the other hand, the same physical system is involved, and from this viewpoint it is the same observer, which is in different states for different elements of the superposition (i.e., has had different experiences in the separate elements of the superposition)."

And the theory does not allow any way of deciding which of the different observer states is the "true" one. "It is . . . improper to attribute any less validity or 'reality' to any element of a superposition than any other element," Everett wrote. "All elements of a superposition must be regarded as simultaneously existing."

Chapter 10

From Black Holes to
Supermatter

We can, of course, never be entirely certain that Ohm's law holds on the dark side of the moon, unless we have been there to test it. In the same way, we can allow continued speculation that quantum mechanics will fail somewhere. I would not want to bet my own money on that.

—ROLF LANDAUER, 1989
"Computation, Measurement, Communication and Energy Dissipation"

Nothing in astrophysics captures the imagination quite like a black hole. After all, black holes capture everything else. Even the name itself captured physicists' attention back in the days when not many people believed that black holes really existed. "The advent of the term black hole in 1967 was terminologically trivial but psychologically powerful," wrote John Wheeler, the man credited with conceiving that name. "After the name was introduced, more and more astronomers and astrophysicists came to appreciate that black holes might not be a figment of the imagination but astronomical objects worth spending time and money to seek."[1]

When I was a student in Wheeler's physics class at the University of Texas, he mentioned casually one day that he had named black holes, but I was too dumb then to inquire about the story behind it. For two more decades, despite numerous opportunities, I never thought to ask him how he had come up with that name. But in February 1998, when I went to Princeton to interview him about the physics of the twentieth century, black holes were at the top of my question list.

Wheeler, eighty-seven years old at the time, talked softly and slowly, although not much more softly or slowly than he had twenty years earlier. I always had the impression that a lot of thought preceded each syllable. This time there was an additional complication in our communication, though—he had lost his hearing aid. That made it hard to ask subtle questions. But I discovered a useful strategy—I just needed to shout a phrase loud enough to get him started. I mentioned "H-BOMB" and he would talk at length about his work on the hydrogen bomb project. And then "BLACK HOLES" got him started on the black hole story.

It was 1967, he said, and pulsars had just been discovered by Jocelyn Bell in England, earning her adviser, Anthony Hewish, a Nobel Prize. The discovery of these interstellar lighthouse beacons immediately started theorists scrambling to figure out what they were. (It turns out, of course, that they are neutron stars, dense balls left over from supernova explosions, with the mass of a star crushed into the size of a city.) Anyway, in the fall of 1967 a conference in New York was convened to discuss this new phenomenon.

Wheeler had been studying the fate of massive stars, and he believed that some of them would explode and leave behind a core so massive that its own gravity would crush it into nothingness. Instead of a neutron star, such an event would leave behind an intensely strong gravitational field surrounding a "gravitationally completely collapsed object." In his talk at the New York conference, Wheeler repeatedly referred to this "gravitationally completely collapsed object." Eventually one of his listeners couldn't take it any more.

"Somebody in the audience piped up, 'Why not call it a black hole?'" Wheeler recalled.[2]

He still doesn't know who suggested it, but Wheeler liked the idea and used "black hole" in a talk he gave in December 1967.[3] When the talk was published in 1968, it marked the first appearance in print of "black hole" to describe a gravitationally completely collapsed object. Wheeler just slipped the new term in while discussing the collapsed core of an exploded star: "What was once the core of a star is no longer visible. The core like the Cheshire cat fades from view. One leaves behind only its grin, the other, only its gravitational attraction. . . . Moreover, light and particles incident from outside emerge and go down the black hole only to add to its mass and increase its gravitational attraction."[4]

Wheeler explained that nothing can escape a black hole's powerful gravity—not even light, which is why black holes are black. Anything that falls into a black hole, stays in. They are the bottomless pits of outer space.[5] Of course, any information contained by anything falling in a black hole is lost as well. And any object of any kind contains some information. It doesn't have to be printed or recorded—any object's structure contains information about how its parts are related. So black holes are the ultimate wastepaper baskets. Any information falling in—whether in a book or a brain—would be gone forever.

To which you might reasonably reply, "So?" But to which a physicist would exclaim "What?!"

The problem for physicists is that information is not supposed to get lost in a universe ruled by quantum mechanics. It should be possible, in principle, to trace back everything that has happened. (In other words, even though you cannot predict the future with quantum mechanics, you should be able to reconstruct the past.)

Physicists don't worry about losing information when a bomb blows up a library. In that case information isn't really lost. Subtle patterns remain in the atoms and photons that fly away. It wouldn't be easy, but in principle all the atoms and photons could be tracked down and measured. Their current conditions would permit a reconstruction of the precise initial conditions before the explosion.

But instead, drop a library into a black hole and physicists start to sweat. All you can measure about a black hole is how much mass it has, how much spin it has, and its electrical charge. From the mass of a black hole you can figure out how big it is. But you can't find out how it got to be that big—whether its mass came from swallowing gas or rocks or chairs or Sarah McLachlan CDs. And you can get nothing that falls into a black hole out again. If you drop a library book into a black hole, it will be forever overdue. It does no good to argue that the information is there, inside the black hole, merely hidden from direct view. Black holes not only swallow information, they chew it up. In addition to making excellent wastepaper baskets, black holes may also be the ultimate shredders.

At least, that would seem to be the situation if Einstein's equations for a black hole's gravity are correct—and most physicists fervently believe that Einstein's equations are correct. But nearly all physicists believe even more fervently that quantum mechanics is correct, too. And quantum mechanics says you're not allowed to lose information. Black holes may be the ultimate dead end on the information superhighway, but quantum mechanics enforces all the traffic laws. Quantum mechanics says information is never lost. The ability to reconstruct the past from the future is guaranteed. If black holes dispense with information as advertised, then the guarantee is violated and quantum mechanics would fail—more astounding than Perry Mason losing a case.

"We certainly do believe in gravity, we do believe in quantum mechanics, and it seems that the two of them combined have led us to a paradox and a puzzle, an alarming inconsistency," says physicist Leonard Susskind. [6]

Black Holes and Bits

It was December 1992 when I heard this paradox posed explicitly by Susskind, a Stanford physicist who was in Dallas to give a talk. Those were the days before Congress had killed the superconducting super collider, the giant atom smasher under construction a little south of Dallas. Susskind came to town to deliver a seminar at the super collider laboratory.

I went to hear Susskind's talk because I'd met him once before, at the Aspen Center for Physics, and found him smart, straightforward, and articulate, my favorite kind of scientist. He is a real "tell it like it is" guy, sort of the Howard Cosell of physics, always offering precise and insightful criticism. And he was clear and direct about what he thought of ideas for solving the black hole information paradox: "All the proposals," he said, "flop on their ear."[7]

Susskind's talk was the first time I had heard such a dramatic concern about the loss of information in the universe. I had forgotten, or never realized, that quantum mechanics requires information about the past to be permanently recorded. But I recalled that Wheeler had stressed to me that black holes swallow information, and that a black hole's size (as measured by the surface area of its horizon) reflected how much information the black hole had swallowed. That, of course, was the whole point of drawing a picture of a black hole with its surface covered by 1s and 0s.

Long before Wheeler drew that picture, he had wondered about the implications of a black hole's ability to swallow information (or entropy, which amounts to the same thing. As Wheeler puts it, entropy is just "unavailable information").[8] The surface area of the black hole's event horizon—that imaginary "point of no return" boundary—reflects the black hole's information content.

"The number of bits that go down the black hole directly gives you the size of the black hole, the area," Wheeler had explained to me back in 1990.[9] That is, the black hole's spacetime and mass are intimately connected with the quantity of information it contains. Just why that is so is an issue at the frontiers of astrophysics today. Exploring this issue, Wheeler believes, is an essential aspect of the quest to understand his notion of It from Bit.*

In one of his books, Wheeler tells a story that clearly foreshadows the information-loss paradox. In a conversation in 1970 with Jacob Bekenstein, then a graduate student, Wheeler remarked on his "concern" about increasing the entropy of the universe by placing a cup of hot tea next to a cup of cold tea. As heat is exchanged—the hot cup cooling, the cold one warming—the total energy of the universe is unchanged. But the tea molecules that had been neatly segregated into hot and cold are now jumbled, causing the universe's disorder, the entropy, to increase. And that increase in entropy cannot be reversed. As Wheeler put it, "the consequences of my crime, Jacob,

---

*Wheeler's standard expression of the notion of "It from Bit" is as a working hypothesis that "every 'it,' every particle, every field of force, even the spacetime continuum itself, derives its function, its meaning, its very existence entirely—even if in some contexts indirectly—from the apparatus-elicited answers to yes or no questions, binary choices, bits."

echo down to the end of time."[10] However, Wheeler noted, suppose a black hole were to pass by. He could toss the teacups into the black hole, and the increase of entropy is concealed from view forever. No one would ever know about the entropy that Wheeler's teacups added to the universe.

Bekenstein mulled this over, Wheeler says, and a few days later announced that Wheeler was wrong. The entropy increase in the universe would not be hidden because black holes possess entropy, Bekenstein had concluded. Tossing in the teacups would increase the black hole's entropy. In fact, tossing anything into a black hole increases its entropy. And as it gobbles up matter and increases its entropy, a black hole also grows bigger. Entropy, therefore, is related to a black hole's size. In fact, Bekenstein calculated, a black hole's surface area grows in direct proportion to the entropy it possesses. That's why the surface area reflects the black hole's information content—remember, information is just another measure of entropy.

Bekenstein went on to write a famous paper describing black holes in thermodynamic terms. (Entropy, after all, is governed by the second law of thermodynamics.) Somehow thermodynamics played a part in understanding black holes, although it wasn't clear exactly how. And then along came Stephen Hawking.

Hawking is no doubt the most famous and most immediately recognized physicist of the era, the nineties' answer to Einstein as a human symbol of science. There is something compelling about a scientist in a wheelchair, unable to talk (except through a computer-voice synthesizer) but still able to think, and think deeply, about the secrets of the cosmos.[11]

In the mid-1970s, Hawking realized that black holes are even more interesting than Wheeler and Bekenstein had imagined, because in fact, black holes are not really completely black, just a very dark shade of gray. This gray area Hawking discovered surrounds the black hole's edge, or event horizon, an invisible boundary at a certain distance from the black hole's center. (The exact distance is determined by the amount of mass collapsed into nothingness at a black hole's center.) A space traveler passing through an event horizon toward the center would feel nothing special at this boundary. (A while after passing through the event horizon, though, such a traveler would get stretched into a submicroscopic spaghetti strand by the irresistible gravitational pull.) In any case, inside the event horizon, nothing can get out. A black hole could emit nothing. Or so conventional wisdom contended.

But Hawking saw a loophole. Space can produce temporary incarnations of tiny subatomic particles. It's just the way space is. It's one of those quantum things—for a brief instant, energy in the vacuum of space can be converted into particles of matter or light. It breaks no rules as long as the particles rapidly convert themselves back into energy. So if the matter particles created are say, an electron and its antimatter counterpart, the positron, they can recombine and annihilate in an eyeblink.

But suppose, argued Hawking, that this particle-pair creation occurred at a black hole's event horizon. Then one of the pair could fall inward, but the other might escape and fly away. To a distant observer, it would appear that the black hole had emitted a particle. Furthermore, the energy needed to create that particle would have to be paid for by a slight reduction in the black hole's mass. You could view this as the infalling particle carrying negative energy, the outgoing particle a balancing amount of positive energy. The addition of negative energy particles to a black hole would diminish its mass.

Nothing special has to happen to make this particle-pair producing process occur. It automatically occurs, in all of space, all the time. So it will happen all the time around any black hole. Consequently black holes, deprived of any new food to swallow to get bigger, will, left on their own, get smaller. They'll lose weight forever as particles fly off to distant space. In other words, black holes evaporate.

On the one hand, that's good news, because it explains a mystery in Bekenstein's analysis of black hole entropy. If black holes have entropy, as Bekenstein asserted, they must have a temperature, since entropy is defined in terms of temperature. But to have a temperature, a body has to emit radiation. Black holes supposedly weren't allowed to emit anything. So Hawking showed how black holes could, in fact, emit radiation after all, and therefore have a temperature. The radiation particles that evaporating black holes emit is now known as Hawking radiation.

Hawking calculated that his radiation would be entirely random—it would contain no patterns, no information related to anything that had fallen into the black hole during its lifetime. And that realization made black hole evaporation cause for some alarm. At first a black hole would evaporate slowly, but the smaller it got, the faster it would evaporate. As it shrunk to subatomic size it would ultimately explode, disappearing in a poof of gamma rays. So black holes are not only wastepaper baskets, but shredders as well. Only in this case, the shredding machine itself disappears, along with the information it has shredded.

As long as black holes stayed stable, information loss posed no particular problem. Sure you couldn't get the information, but it wasn't lost, you knew where it was, it was in that black hole. If you really wanted it that badly, you could jump in after it. But if black holes evaporate and then go poof, the information inside is gone, too.

So what happened to it? In his talk in Dallas, Susskind outlined three possibilities. One was that information really is lost in a black hole, lost forever, and something must be done to modify quantum mechanics. Another proposal contended that the black hole does somehow spit out the information it has swallowed. (Perhaps Hawking radiation did contain some subtle pattern after all that concealed the missing information.) Still others believed that the information falling into a black hole would be packed into a tiny particle left behind after the black hole had evaporated. But in each of these three scenarios, something important was violated—quantum mechanics, Einstein's theory of relativity, or common sense. As Susskind said, "all the proposals flop on their ear."

I found Susskind's talk intriguing, and realized that this was a good story—physicists baffled by black holes, the foundations of physics called into question, greatest minds stumped. It was too good to waste on a mere column; I decided to write a long story, going into the gory details. This came at an ideal time, for I was about to go to Berkeley for the Texas Symposium on Relativistic Astrophysics, where all the leading researchers interested in this issue would be talking.[12] Even Hawking himself would be there. And Frank Wilczek, of the Institute for Advanced Study in Princeton.

Wilczek had given me a paper on the black hole problem when he visited Dallas the previous January. (Had I paid more attention to his paper then, I would have realized what a good story it was months earlier.) So I talked with Wilczek at Berkeley, who said he thought somehow or other black hole physics would have to be modified to save quantum mechanics. He thought that in some way the information could leak out of a black hole. But another physicist speaking at the meeting, John Ellis of CERN (the European atom-smasher laboratory, near Geneva), thought it more likely that the problem was with quantum mechanics. "My view is that we're going to have to modify quantum mechanics," Ellis said at the symposium.

Hawking, on the other hand, contended that the information really does get lost—streaming from the black hole's interior into some other universe, disconnected from the one we all live in except by spacetime umbilical cords called wormholes. "I think the information probably goes off into another universe," Hawking said. "I have not been able to show it yet mathematically, but it seems to be what the physics is telling us. But most other people are rather conservative and don't like the idea of loss of information because it is not what they were taught in graduate school."

At a news conference, I asked Hawking about Ellis's contention that if information really is lost like that, quantum mechanics would have to be changed. "People like Ellis accuse me of wanting to change quantum mechanics," Hawking replied. "That is an exaggeration. I'm just extending quantum mechanics to a situation we have not considered before."[13]

Other leading scientists at the meeting had ideas of their own to explain the information-loss paradox. Gerard 't Hooft, of the University of Utrecht in the Netherlands, suggested that information actually escapes from a black hole as it falls in. Somewhere in the vicinity of the event horizon, an invisible "cusp" could shoot out particles carrying information away as other particles enter, he conjectured. But it was just an idea, and he hadn't worked out any details. "This theory is a long way off from even getting properly started, let alone completed," he said.

Robert Brandenberger of Brown University presented a case for modifying Einstein's general relativity theory to allow black holes to leave behind stable remnant particles after evaporating. All the information ever falling into the parent black hole would be crammed into the leftover baby particle, forming something like a fantastic computer memory chip. But it wasn't obvious to everybody that some tiny remnant could contain all the information a black hole had swallowed. After all, a black hole could suck up a lot of information in its lifetime.

But Susskind speculated that the interior of a tiny remnant particle might contain "balloons" of information storage space that are hidden from outside viewing. To me that sounded like a "Dr. Who" approach, after the British TV science-fiction character. Dr. Who travels in a time machine that from the outside looks to be the size of a telephone booth, but on the inside, it's full of spacious rooms and laboratories.

Wilczek, on the other hand, preferred the idea of information escaping from a black hole before it evaporated entirely away. It's possible, Wilczek said, that as a black hole shrinks, it might reach a point where radiation can escape and carry information away with it. His idea did require certain changes in standard black hole physics, though.[14]

Another new approach to the problem was proposed by Susskind a few months after the Berkeley meeting. He noted that the paradox of black holes and information was similar to the situation in the 1920s, when physicists were baffled by experiments proving that electrons are waves and that light is made of particles, whereas earlier experiments seemed conclusive that light is a wave and the electron is a particle. Niels Bohr ultimately solved the wave-particle mystery with his principle of complementarity (an electron can be either a particle or a wave, depending on how you look at it). In 1993, Susskind suggested a possible solution to the black hole information problem by borrowing a page from Bohr's book. The information is swallowed by a black hole or it isn't, Susskind said, depending on where you are when you're looking at it.

To an observer falling into a black hole along with an information-bearing particle, the information goes on down the drain, along with the observer, who is soon to be stretched into a human piece of spaghetti near the black hole's center. But to an observer safely viewing the particle from a distance, the information spreads out around the black hole and floats away, Susskind told me. "Things get extremely distorted near the horizon, so different observers see different versions of what takes place," he said. "It's not two events, it's two different observations of the same event."[15]

In other words, the location of information gets complicated.

Observers in different places simply won't agree on where the information is. "Either you're outside or you're inside," said Susskind. "You shouldn't try to think of physics in terms of a super-observer who can see inside and outside the black hole at the same time." He calls this explanation the principle of black hole complementarity, after Bohr's use of complementarity to describe wave-particle duality.

Information and Gravity

At the time of the Berkeley meeting, it was obvious that nobody really knew how to solve the black hole information paradox. It was equally clear where the root of the problem was. Black holes were all about gravity; information preservation was all about quantum mechanics. And the great failure among the magnificent accomplishments of twentieth-century physics was its inability to find a theory that unified quantum mechanics with gravity.

Nowhere was that failure more clearly exposed than with the black hole information paradox. And therein lies one of the most fascinating conjunctions of different lines of thought in modern science. Here in black holes, at the intersection of gravity and quantum mechanics, was another instance of information playing a central role at the scientific frontier. Until then, the emphasis on information in physics had been mostly on the fringe. It was hardly mainstream science. The information-loss paradox, however, was smack in the middle of the cosmological stream. It was an important issue, a never-ending crisis for quantum mechanics and general relativity. It suggested that understanding the tension between those pillars of modern physics would first require a better understanding of—information!

This raised the possibility that information theory and the physics of computation could have some real relevance to astrophysics. And in fact, at the 1992 physics of computation meeting in Dallas (the meeting where Ben Schumacher coined the term *qubit*), Chris Fuchs (rhymes with books) applied Landauer's principle to analyze black holes as information erasers. Landauer's principle says that erasing information uses up energy, and using up energy means increasing entropy. If information erasure is what's going on in a black hole, then Landauer's principle should apply. If so, Fuchs reasoned, it ought to be possible to calculate the relationship between loss of bits and the resulting increase in the black hole's surface area. The Bekenstein-Hawking formula says a black hole's entropy should be equal to one-fourth the surface area (when measured in appropriate units). Using some approximations and assumptions, Fuchs calculated that according to Landauer's principle, the entropy as a fraction of the surface area had to be at least 0.18, not so far off from one-fourth (0.25). Considering that Landauer's principle was devised for computers, far removed from the realm of black holes, this was a remarkable agreement, suggesting that Landauer's principle "actually has some predictive power," Fuchs said at the meeting.[16] "This result argues well that . . . Landauer's principle is valid outside its first arena of justification," he wrote in the conference proceedings.[17]

To me it suggests that Landauer's principle (and by implication, the notion that information is physical) has a deeper significance than physicists generally yet realize. And that the physics of information and computation may offer important insights into real-world cutting-edge scientific problems—such as finding a theory of quantum gravity. Or in other words, reconciling quantum mechanics with general relativity.

General Relativity

Black holes embody the essence of general relativity theory. Just as they capture information, and physicists' imaginations, black holes capture the most basic features of space and time and gravity, features that are best described by Einstein's 1915 theory of general relativity. As a theory of gravity, general relativity represented a radical departure from the traditional gravity of Newton. Newton's law of gravity is summed up by saying "what goes up, must come down." In Einstein's gravity, the proper slogan would be more like "What goes around, comes around." In other words, gravity is geometry.

Now, "gravity is geometry" has a nice ring to it, but it needs a little more discussion. Like, for example, what does it mean, the geometry of spacetime? For more than two thousand years, everybody had been taught geometry from the viewpoint of Euclid, a Greek geometer and a very good one. He thought space was flat. In other words, space was like the surface of a sheet of paper—parallel lines never meet, the angles of a triangle add up to 180 degrees, all those other things you learned in geometry class. But those things are right only if space is flat like a piece of paper, not curved like the surface of a sphere. On a sphere, like the Earth, longitude lines are parallel, but they do meet, at the poles. And the angles of a triangle add up to more than 180 degrees.

Until the nineteenth century, most scientists and mathematicians (and especially some philosophers) believed that Euclid's geometry was the only possible geometry. Space could not be shaped in any other way. But then some free-thinking geometers saw ways to construct alternatives to Euclid, working out sets of axioms that described geometries in which space was not flat, but curved. Some began to wonder whether real space was really like the flat space of Euclid or whether it had a more complicated geometry. Did a ray of light really travel in a straight line? If you measured the angles between three distant mountains, would they really add up to 180 degrees? (The mathematician Karl Friedrich Gauss, incidentally, allegedly attempted this experiment. The historical evidence suggests that he did attempt to measure the angles between three mountains but not for the purpose of testing Euclidean geometry. He was investigating whether surveying measurements that assumed the Earth to be a perfect sphere were sufficiently accurate.)[18]

When Einstein was working on describing gravity, he found that the version of non-Euclidean geometry developed by Georg Friedrich Bernhard Riemann was just what the doctor ordered. In Einstein's general relativity, space has a nonflat geometry whenever mass is around. Matter warps space, causing it to curve. Matter moving through space follows the curvature of space. In the words of John Wheeler, mass grips space, telling it how to curve; space grips mass, telling it how to move. In other words, planets orbit the sun and rocks fall to Earth not because of some mysterious pull. Rather, objects merely move through space on paths determined by the geometry of that space. Rocks fall to Earth not because of a downward tug from the ground, but because the ground gets in the way of the rock's natural trajectory.

Technically, and it's an important technicality, gravity is really the geometry not merely of space, but of the combination of space and time, or spacetime. This was an essential part of relativity's success. Riemann himself had tried to describe gravity by the geometry of space but failed, because he limited his analysis to space's three dimensions. He didn't know he needed to include time as a fourth dimension.

Fortunately, Einstein knew better, because time as a fourth dimension was already an established part of relativity theory. General relativity was the second of Einstein's two theories of relativity. The special theory came first—Einstein called it the special theory because it was his first relativity theory and the first one is always more special than the rest. Not really. It was "special" because it described only uniform motion—that is, motion at a constant speed in a straight line. In special relativity, Einstein sought to show how the laws of motion would remain the same for observers moving (with respect to one another) at a uniform rate. (General relativity "generalized" the special theory by applying the same idea to accelerated motion. In general relativity, accelerated motion is equivalent to the motion of a body in a gravitational field.)

It turned out in special relativity that making the laws of motion stay the same had some extraordinary consequences. For one thing, no matter how fast anybody moved, their measurement of the speed of light would always come out the same. For another, clocks in a fast-flying spaceship would appear to run slowly from the point of view of an observer on the ground. So a space traveler might return from a two-year trip to find that all his friends had aged by decades.[19]

For the math to work out consistently, special relativity had to merge space with time, making time the "fourth dimension." (Actually, it was one of Einstein's professors, Hermann Minkowski, who showed how special relativity combined time with space to make spacetime.) So Einstein knew about the need to do that when working on his general theory.

(An interesting historical note is that in 1885 a letter to the editor of the British scientific journal *Nature* proposed that time should be considered the "fourth dimension." The author of this letter, identified only as "S," pointed out that time couldn't be added to space very easily, so that it would be necessary to describe a geometry with time and space combined. And he chose a brilliant name for this combined space and time, namely "time-space." Close, but no Nobel

Prize for that. If only he had thought to combine them in alphabetical order.)

Relativity as Symmetry

The main point about relativity is this—Einstein invented it as a way of having a theory that described the laws of nature without regard to how anybody was moving. It would be an inconvenient science if the laws changed depending on whether you were standing still, driving a car, or flying in an airplane. The laws of physics have to stay the same from different points of view—in other words, the laws of nature are symmetric. Deep down, the essential message of relativity is the importance of describing nature with a sense of symmetry.

In fact, one of the hallmarks of twentieth-century physics has been the pursuit of fundamental laws of nature through appeal to symmetry—in the words of the mathematician Hermann Weyl, "a vast subject, significant in art and nature." In painting and architecture, for example, symmetry is a supreme consideration. In biology, symmetry describes features simple and elaborate, from a butterfly's wings to people having arms, legs, eyes, and ears on both sides of their bodies.

"Symmetry, as wide or as narrow as you may define its meaning, is one idea by which man through the ages has tried to comprehend and create order, beauty, and perfection," wrote Weyl.[20] And during the last century, science has found that deep notions of symmetry underlie the laws of nature themselves.

The core idea of symmetry is that when something changes, something else remains the same. If a circle is flipped over, or a sphere is rotated, the description of the circle or sphere is unchanged. A baseball park is symmetric if the distances to the left field fence at each point are the same as the distances to the right field fence. In other words, switching left with right leaves the outfield dimensions the same as they were before.

In physics, symmetry principles apply in a more abstract way to the laws of nature. Mathematical symmetries express the unchanging regularities hidden within natural law. For example, changing position in space or in time leaves the laws of nature unchanged.

Otherwise, repeating experiments at different times or in different places would yield inconsistent results, and making sense of the natural world would be impossible.

Of course, all the universe is not totally symmetric. If so, it would be a boring place. In real life the underlying symmetries of nature get "broken." The universe may have begun as a symmetric fireball, but as it expanded and cooled new features emerged, the way drops of water and chunks of ice eventually appear as steam cools and condenses.

Einstein's relativity theories revealed the stunning power of analyzing the symmetry of motion through space. Special relativity says that uniform motion doesn't change the laws of motion, and that symmetry requirement has some bizarre consequences. But as all the experimental tests of relativity have shown, they are necessary consequences. In Einstein's general theory of relativity, the symmetry is even grander. The laws of motion and gravity are preserved even in the face of accelerated motions—acceleration is no different from gravity. It doesn't matter how you are moving or where you are in space and time—the laws stay the same.

In other words, Einstein's theories of relativity describe the symmetry of space throughout the universe. But physicists nowadays suspect that Einstein did not tell the whole story. It may be that by exploring new and grander symmetries, the impasse between merging quantum mechanics and gravity might be eliminated. In fact, scientists now suspect that in addition to the ordinary dimensions of space described by relativity, there might be microscopic "quantum" dimensions of space. If so, there is more symmetry to the universe than Einstein realized. Expressing the underlying sameness of nature that includes these new dimensions will require a new theory of symmetry, known by physicists as "supersymmetry."

Supermatter

"Supersymmetry"—or SUSY—is a mathematical way of relating the apparently different appearances of matter and forces in the universe. So far there is no firm evidence that supersymmetry is real, although there are some hints. But if it is real, it has some amazing implications—the sexiest being the hidden existence of supermatter. If supersymmetry rules the universe, then nature is concealing a companion particle for every basic particle now known. These "superpartner" particles could, in principle, be produced in sufficiently powerful atom smashers. "I think the discovery of supersymmetry," says Caltech physicist John Schwarz, "would be more profound than life on Mars."[21]

That would make it a Page One story, of course. But the truth is supersymmetry may already have been discovered. If superparticles lurk in the cosmos, they may make up the mysterious dark matter that astronomers infer from the way galaxies spin and cluster. So it might not require an atom smasher to make a superparticle—it might be possible to snag one flying through the solar system.

In fact, several labs around the world have been attempting to do just that. One is in Italy, at the underground Gran Sasso laboratory where a collaboration known as DAMA (for dark matter) has sought signs of dark matter particles known as WIMPs (for weakly interacting massive particles). It is a good bet (though not a certainty) that if WIMPs exist, they are superparticles.

In December 1998, I went to the Texas astrophysics symposium, this time in Paris, where the DAMA team reported an intriguing sign of a possible WIMP sighting. The DAMA experiment uses chunks of sodium iodide that should give off a flash of light when a WIMP strikes. Of course, the chunks give off light when a lot of other things strike, too, making it hard to prove that any one flash of light is a WIMP. The obvious solution is the June–December approach. In June, the Earth is orbiting the sun in the same direction as the sun is speeding through the galaxy. In December, the Earth moves the opposite direction. If the galaxy is full of WIMPs, more should hit Earth as it heads along with the sun into the WIMP wind than when it's moving away. It's sort of like the way your car windshield smashes into more bugs going forward than when going in reverse.

At the Paris meeting, Pierluigi Belli of the DAMA team said the experiment saw a stronger signal in June than December two years in a row. The team's analysis suggested that the signal was not likely due to chance. It might very well be due to WIMPs, and possibly could be the first detection of a supersymmetric particle.

This discovery did not make big headlines. For one thing, as far as I could tell I was the only U.S. newspaper reporter there.[22] And scientists from other teams criticized the DAMA analysis of the data rather sharply. Other experts I talked to said it was too soon to say that WIMPs had been definitely discovered. But the hint of a WIMP was still exciting. Its estimated mass fell smack in the range expected by theorists who worship supersymmetry.

If the DAMA WIMP does turn out to be a superparticle, confirming supersymmetry, the twenty-first century will explore a very new universe, much as the twentieth century encountered a universe very different from what science knew in the centuries before Einstein.

"Supersymmetry is really the modern version of relativity," says physicist Edward Witten, of the Institute for Advanced Study. "It's the beginning of the quantum story of spacetime."[23]

Chapter 11

The Magical Mystery Theory

Although the assumptions which were very properly made
by the ancient geometers are practically exact . . . the truth
of them for very much larger things, or very much smaller
things, or parts of space which are at present beyond our
reach, is a matter to be decided by experiment, when its
powers are considerably increased.

—William K. Clifford,
The Common Sense of the Exact Sciences

Space, as every *Star Trek* fan knows, is the final frontier.

It's the frontier of exploration. It's the frontier of discovery. It's
the frontier of understanding the very nature of the universe. If Vince
Lombardi were alive today, he would probably say that space is not
really the final frontier—it's the only frontier.

Space (or "spacetime," to be more precise in the era of relativity)*
is where everything about reality comes together. At the frontiers of the

_____

*On occasion I will retreat a little from relativity's insistence on spacetime and
sometimes talk about space and time distinctly. Most of what I say about gravity and geometry will really be about spacetime, but there are situations where
it is easier to discuss space itself. In those cases it is better to compute how space

213

scientific understanding of space and time, scientists have only a few clues, and no certainty, about how to answer the deepest questions. The standard physical ideas of space and time are not just written in books and accepted as established, but are topics of active theoretical interest these days. A lot of smart people are applying high-powered physics and math to figuring out what space and time are *really* like.

This research is producing some very new ideas and theoretical results that are not yet in any textbook. Some of them will never be, because some of them are surely wrong. It's just too soon to tell which of these ideas will lead the way to a new understanding of the universe. Nevertheless, a journey to the frontiers of spacetime research is sure to lead to new signs of the importance of information in the nature of reality and existence.

In a way there's a curious contradiction here. On the one hand, space is synonymous with nothingness. On the other hand, the math describing space is at the heart of theories that attempt to describe everything. I guess it's just another way of saying that somehow the whole universe was created from nothingness, creation ex nihilo, so understanding the universe really means understanding space itself. All the other things that make up a universe, like matter and energy—and information—in some way derive their existence from the "nothingness" of space.

Of course, physicists have known since Einstein that space is not the nothingness of common sense, but a real participant in the fundamental phenomena of nature. In the purest sense, Einstein showed, space is responsible for bringing down whatever goes up, for the twirling of planets around the sun, and for the cluttered array of galaxies spanning the universe. In other words, gravity is geometry—the geometry of spacetime. But even Einstein did not imagine the bizarre twists that the study of space has taken lately. In the last few years, physicists have reported some sudden and surprising advances in figuring out what space is really like. The new results bear on ancient questions—like whether there is a smallest possible distance—as well as on modern efforts to find the ultimate theory unifying all of nature's forces with matter. "It's clear that the nature of space is

---

changes with time rather than describing spacetime as a whole. And at times I will discuss aspects of time itself, apart from space. So please don't get confused.

rather different than how we are accustomed to thinking of it," says theoretical physicist Andy Strominger of Harvard University.[1]

That "accustomed way" of thinking was shaped early in the century by Einstein's general relativity theory. Einstein's description of gravity as the geometry of spacetime produced some marvelous predictions—particularly the way light from a distant star would be bent when passing by the sun. (Newtonian gravity did, in fact, predict that light from a star would curve when passing by the sun, but by only half as much as Einstein's theory predicted. And when the measurements were made, Einstein got the right answer.) Other scientists worked out implications of Einstein's theory to predict other bizarre things, like the expansion of the universe, and black holes. There seemed to be great benefit and beauty in describing the gravitational force as simply the geometry of spacetime.

Einstein's space is a very different space from the absolute space that Isaac Newton talked about, a stage for the scenes of physics to play out. Newton thought of space as an absolute frame of reference in which objects moved. Gottfried von Leibniz, the contemporary and intellectual adversary of Newton, thought otherwise. Leibniz regarded space not as a stage or arena, but literally nothing, just a pattern of relationships among objects—sort of like the branches in a genealogical family tree. The pattern shows relationships but has no physical reality of its own.

Relativity showed that there really is no absolute Newtonian framework of space, so in some sense the space of relativity supports Leibniz's view that space is really nothing. Yet in a different sense it is more something than ever. It's not just an empty receptacle for matter and energy and forces. Matter and energy are intrinsic parts of space. A particle of matter is really a knotted up piece of space. You can think of it as a particle of matter causing a warp in space, but when you get right down to it, the warping and the particle are really the same thing.

And what about forces? In the modern view, force is transmitted by the exchange of particles. A force is the same thing as an interaction, and the way particles interact is to trade other particles. One electron interacts with another, for example, by emitting a photon that the other electron absorbs.

All this raises a deep question: If particles are knots in space, then why aren't all particles alike? The answer complicates space a little. Modern physics says space is full of different "fields." A particle is a knot in a field. A particle's identity depends on what kind of field it is a knot in.

But what's a field? Simply something sitting in space, inseparable from it, like the familiar electromagnetic field that broadcasts radio and TV and cellular phone signals. Photons are knots in the electromagnetic field, just as the matter particles known as quarks are knots in "quark" fields.

Fields make space more complicated than it seemed in the days of Newton and Leibniz. Someday, no doubt, it will turn out to be still more complicated. It remains to be explained, for example, how all the fields in space fit together—or in other words, how they are "unified." Einstein vainly sought a unified field theory that would have combined gravity and electromagnetism. Nowadays physicists speak of "grand unified theories" that combine the electromagnetic fields and nuclear force fields. Getting the gravitational field (that is, the underlying spacetime itself) to join in this unification has been difficult. This seems to be because the fields other than gravity are described by quantum mechanics. And marrying quantum mechanics and general relativity has been harder than negotiating peace in Bosnia.

This is all very perplexing, because quantum mechanics and general relativity are, in their separate domains, the most spectacularly successful theories in the history of science. They've explained a vast range of phenomena already known, and predicted bizarre new phenomena that have been confirmed by experiment and observation.

Quantum mechanics, for example, is at the heart of the "standard model" of particle physics, which succeeds in describing the fundamental particles of nature and the forces that guide them—except gravity. In physics, the standard model is the gold standard of theories.

"This theory is really amazing," says Nathan Seiberg, a theorist at the Institute for Advanced Study. "It's logically self consistent, so whenever we ask a question we get an unambiguous answer—it's always the same answer regardless of how we do the calculation. This sounds like a triviality but it's absolutely crucial. . . . The second thing is that it agrees in a spectacular way with experiment. . . . So this is more or less a complete story, and it's quite beautiful, it agrees with everything. But one thing is left out, and this is gravity. . . . We can't put the two stories together."[2]

So the situation as it stands is something like this. Matter is described in phenomenally precise detail by quantum mechanics. Relativity melded gravity, space, and matter. The frontier of understanding space today is the search for a theory of quantum gravity. On this much everybody agrees. And I think there's general agreement that finding a theory of quantum gravity will involve a deeper understanding of the nature of spacetime. The secrets of space seem tied up not only in the knots of the particles and forces, but also in the universe on the large scale. I think many of the important research issues in this regard can be summarized by discussing the efforts to answer three essential questions about space:

1. What is space like on very large scales?
2. What is space like on very small scales?
3. How many dimensions does space have?

Space on Large Scales

The question of what space is like on very large scales has been in the news a lot in the past few years—largely because the Hubble Space Telescope is such a news maker, and much of what it finds out is relevant to the big questions about the universe. Those questions include, of course, whether space on the large scale is curved so much that it will eventually collapse and we'll have a big crunch, or whether it will keep expanding forever.[3] We don't know yet for sure, although at the moment the best bet is no crunch. In fact, the universe may very well be expanding at an ever accelerating rate. That evidence is not conclusive, though, and there could always be more surprises about what the universe is actually doing.

There are other questions about space on large scales that current astronomy is trying to answer. For example, how does the large-scale geometry of space affect things like the development of clusters of galaxies? Usually this is believed to be merely a question of general relativity. Relativity describes the universe on large scales, while quantum physics describes small scales. But there are some new analyses that suggest quantum effects might have a role to play on large-scale geometry.[4]

Space on Small Scales

Space on very small scales, and I mean really very small scales, is a big preoccupation of physicists these days. They want to know what happens to the laws of physics when distances under consideration are much smaller than the smallest of subatomic particles. Einstein's equations do a good job of describing space, but at very tiny distances, about 10 to the minus 33 centimeters, problems develop. This distance is called the Planck length, and it's very, very small—enlarging something the size of the Planck length to the size of an atom would be like making a football stadium bigger than the entire universe.

So if you squeeze matter into a tiny space, it gets denser and denser, and space curls up very tightly. If you keep going and space curls up below the Planck length, Einstein's equations become helpless. They can no longer describe what's going on. This is what they say happens inside a black hole, at the center, where the density of matter is infinite, all the matter crushed into a single point called a singularity. There is no way to describe a singularity—space and time literally go away there—and so many physicists hope that singularities don't really exist.

The only way to figure out what's going on at distances so short, most experts believe, is to come up with a theory that combines quantum mechanics with general relativity to provide a complete quantum theory of gravity that will describe everything just fine. It's not an easy assignment. People working on this problem have come up with a lot of suggestions about what might happen at the Planck scale. Some suggest that there is just a shortest possible length. In fact, most attempts to combine quantum physics with gravity reach that conclusion. Just how that shortest possible length manifests itself physically is another question. Some theories suggest that if you try to cram more matter (or energy) into such a small space it doesn't curve the space up even tighter, it starts to make things bigger. Inside a black hole, what might happen is space that has been compressed down nearly out of existence bounces back and starts expanding, possibly making a whole new universe.

But there are other attempts to describe what space is like physically at these small scales, even without matter and energy around—in other words, a true vacuum. Quantum theory seems to require that it must be pretty turbulent down there. The vacuum of space is not a boring nothingness, but a lively place. Because of the uncertainty principle, you aren't allowed to say that there is zero energy in empty space. That would be too certain. Some energy can show up here and there for a short time; as long as it disappears later, the quantum rules are satisfied. Long ago, John Wheeler suggested, therefore, that spacetime is a "foam" with little tunnels popping into existence that lead from one part of space to another. These tunnels are called wormholes.

The study of wormholes has become quite elaborate in the decades since Wheeler's original suggestion. Several physicists have proposed that space could be laced with wormholes that serve as passageways to entirely different universes. Stephen Hawking has been a major contributor to wormhole theory. But lately he has decided that space might not really behave this way. In fact, he has written papers suggesting that rather than tiny wormholes, tiny black holes pop into existence throughout space at the scale of the Planck length. These black holes are like minuscule bubbles that appear in pairs and then disappear. Hawking calls this idea the "quantum bubbles picture." He first had this idea in the late 1970s, before the wormhole picture came into vogue. But he couldn't work out the math for a black hole to appear out of nothingness and then disappear again, as the quantum rules ought to dictate. Eventually, though, he realized that the math works if the black holes appear in pairs.

During the time that a temporary (or "virtual") black hole exists, it could interact with other particles in the universe. For example, a particle might fall into a virtual black hole and then reemerge, but as a different particle, Hawking has suggested.

This would be another example of a black hole erasing information, causing worry among some physicists. But not Hawking. He thinks the black-hole-bubbles picture of small-scale space could have its advantages. In particular, he thinks it could lead to predictions of certain features of particle physics, like details of the strong nuclear force and the nature of the Higgs boson, a particle being sought today in powerful atom smashers. The bubble picture shows, in other words, how the consequences of space's structure on the smallest of scales could be observed in experiments within the reach of physical instrumentation. And that's just the sort of thing that a quantum gravity theory needs to be taken seriously.

On the other hand, it is far from a sure bet that Hawking is right about any of this. There are many other approaches to understanding space on small scales, based on Einstein's relativity but going much farther. One approach in particular that interests many physicists is called the loop representation of quantum gravity.

Loopy Gravity

These loops are not a form of fruity cereal, but mathematical rings related to physical force. The theory basically suggests that space itself is made of these loops. The loop approach surfaced in the mid-1980s when the mathematical physicist Abhay Ashtekar, now at Penn State, discovered a new and simpler way to write Einstein's equations for general relativity. Ashtekar's version used the mathematical form of a force field.

The underlying idea had a much earlier origin, beginning with Michael Faraday, the nineteenth-century British physicist who pictured magnetic and electric fields as networks of lines of force. Those lines could make loops from one magnetic pole (or electrically charged particle) to another. In modern theories of force fields, a line of force in the absence of particles or charges circles back on itself, forming a closed loop. Similar loops appear in Ashtekar's mathematical description of general relativity. And since the loops could be regarded as an aspect of the quantum properties of a force field, Ashtekar and others began exploring this approach to Einstein's theory as a way of connecting general relativity with quantum mechanics.

In 1995, for example, Lee Smolin of Penn State and Carlo Rovelli of the University of Pittsburgh (he has since moved to France) showed how the geometry of space can be determined by how

Ashtekar's loops link up. As a consequence of how these loops link, space cannot come in any size you want, but only in certain allowed amounts. It's as if space could come in units the size of golf balls, tennis balls, and basketballs, but nothing in between, much the way electrons in atoms can possess only certain amounts of energy. In other words, space, like energy and matter, also comes in quanta, like pennies and nickels, but no half pennies or half nickels.

It just so happened that at the time Smolin and Rovelli's paper came out, I was visiting the Institute for Advanced Study in Princeton, as was Smolin. So I went to talk to him about what was going on. Smolin is exceptionally polite and soft-spoken for a physicist, and a deep and sensitive thinker. He was very helpful in explaining both the physical and philosophical aspects of his work.

Of major significance, he said, was that he and Rovelli did not start with the assumption that space has a short-scale foamy structure. This quantum nature of space emerged from their calculations about how loops link. It was very much as Leibniz had suggested—space was all about relationships, not points.

"Points of space are meaningless," Smolin insisted. "The only way to define spatial quantities is by giving relationships between things. The loops themselves are going to define the geometry of space. So the only thing that should matter is the relation between them—how they intersect, how they're knotted together and linked together in three-dimensional space."[5]

Using this approach, Smolin and Rovelli had shown how the geometry of space can be described by the way these quantum loops are linked or knotted together. Since the geometry of space can change over time, the loops must be able to link in different ways. And as it turned out, different loop patterns correspond to structures called "spin networks."

Spin networks had been invented in the 1960s by Roger Penrose, the mathematician who nowadays worries more about consciousness. They can be depicted as a grid of lines that meet at points, called vertices or nodes. In Ashtekar's gravity, the loops can be viewed as connected in a spin network; the loops come together at nodes that represent a small volume of space.

Such a spin network also provides natural spots where particles of matter fit in nicely. And the network provides perfect places for intersections with loops from other force fields, such as the nuclear forces. Thus the loop representation could provide the framework for understanding all of nature's particles and forces, although Smolin suggests that loops might not be able to do the job alone. That task, it appears to many physicists, is a job for superstrings!

Supercontroversy

Superstrings sound like what you might call an all-star team of violinists, or maybe a new high-tech product to replace twine. But in fact, superstrings are unimaginably tiny objects that may be the basic building blocks of everything. And superstring theory is the most popular approach among physicists today for unifying nature's forces and solving the problem of quantum gravity.

At this point I have to inject a rather perplexing aspect of all this research. Despite the fact that the loop quantum gravity approach is so interesting, most of the more prominent physicists interested in quantum gravity prefer superstrings. In fact, many of them dismiss the loop approach rather contemptuously. "The big challenge is to find a coherent theory of quantum gravity," a prominent theorist recently told me. "String theory is the only candidate for that that we have at the moment."

"What about loop quantum gravity?" I asked.

"String theory is the only candidate that we have at the moment," he replied.

Others are more likely just to say that string theory is the best approach to quantum gravity. But in any case many string theorists seem to regard the loop approach as, well, loopy.

This attitude baffles me a little. Some string theorists will say they feel this way because the loop approach has not really accomplished anything significant. But other physicists tell me that the dispute runs deeper and is colored by some aspects of scientific culture. In particular, the loop theorists generally have an intellectual genealogy rooted in general relativity. The string fans tend to be particle physicists. To me it's intriguing and instructive to realize that the way scientists view new research approaches can be shaped so significantly by their intellectual history, even within the discipline of physics.

It may well be that the string theorists are right and the loop approach will never get anywhere. Personally, I plan to suspend judgment and continue to pay attention to both camps. In the meantime, it is certainly true that superstrings have provided the most excitement in the quantum gravity world.

Superstrings are difficult to describe. They are typically very small—down near the Planck-length size. And they are one-dimensional. That is, they are like a line (or a string), rather than like a point. They come in various versions. In some cases they would best be described as loose snippets of string; in other versions they are connected end-to-end to make loops (of an entirely different kind from the loops of the loop quantum gravity approach).

Superstrings became popular in 1984 when John Schwarz of Caltech and the British physicist Michael Green showed how string theory avoided some of the mathematical paradoxes plaguing other attempts to unify quantum mechanics with gravity. In standard physics, nature's basic particles, such as quarks and electrons, are considered to be tiny points of mass, with zero dimension (in other words, no spatial extension). They were the physical version of the fundamental point in Euclidean geometry. But in superstring theory, point particles instead become tiny loops of vibrating string. (The loops are one-dimensional, like Euclidean lines.) Different particles correspond to different modes of vibration, just as a violin string can play different musical notes. Superstrings presumably could vibrate in any allowable way; therefore they very well could be responsible for all the basic particles known in nature.

Besides describing all the basic particles, string theory naturally wraps gravity into the same mathematical package with the electromagnetic and nuclear forces. In fact, the string math requires the existence of a peculiar particle that physicists soon realized was just the graviton, the particle form of gravitation. Many physicists therefore believed that string theory could in principle describe everything in the universe—hence its common label of a "theory of everything."

But superstring theory posed one serious problem. It raised an ugly question about how many dimensions space has.

Remember back for a moment to the early 1900s and how hard a time people had grasping relativity's requirement that there were four dimensions instead of three. It seemed strange, but it was the way to make the math work. In a similar way, superstring math works only if space has a whole bunch of extra dimensions. The number differs in the various versions of superstring theory, but in the most popular version you need ten dimensions of space and time (nine of space, one of time) to get the math to work out right.

Astute critics of this idea immediately asked an embarrassing question. If there are six extra dimensions of space, where are they? Ever since scientists had thought about it, it seemed obvious that space was three-dimensional—that is, you could always specify your position precisely with just three numbers: latitude, longitude, and altitude, if you were doing so in reference to the Earth. You could likewise describe any change in your position by an amount of movement in just three possible directions. So where would any other dimensions be? The answer turns out to be pretty simple—they are everywhere. They're just too small to notice.

This does seem to be a rather difficult explanation to grasp. How can a dimension of space be so small? When trying to explain this once to a friend at dinner, I stumbled onto an idea that I think conveys the right idea. If you go to the North Pole, all of a sudden you find you have lost a dimension. If you take a step in any direction, it is to the south. In other words, you can change your latitude, but you have no longitude to change. But now imagine yourself as an ant, or better yet, a subatomic-sized creature vastly smaller than an ant. You can step the tiniest distance away from the North Pole and now be at a specific longitude. But at a sufficiently close point to the pole, you could circle all the way through all the possible longitudes in an instant. You would be moving through a dimension, yet you could move through it and back to your starting place so quickly that you might not even realize that you had moved.

I've heard some physicists try to make the same point using a straw. Consider an ant crawling along a straw, a very narrow straw. To the ant, there's only one dimension here, a straight line to move along, forward or backward. The ant doesn't realize there's another dimension—that he could take a left turn and walk around the straw. And the point is this—even if the ant turned left and walked around the straw, he'd get right back to where he had been immediately, if the straw is narrow enough. In other words, if that extra dimension is really small, the ant would return to his starting point so fast that he would not even notice that he had moved in another dimension.

So, in the superstring picture of the world, it seems that we may live in a space with nine dimensions. We notice three of them. In the other six, space is curled up so tightly that nobody notices them. If we could move through the other six, we'd get back to where we started from basically instantaneously, so we wouldn't notice anything fishy.

Superstring physicists assumed from the outset that these extra dimensions were very small, possibly because back at the big bang, when the regular three dimensions started expanding, the others just stayed small. And from a mathematical point of view, extra dimensions are no big deal, if you can find the right math to describe how they curl up.

But that presented another problem. In the equations describing the extra six dimensions, there are thousands of different possible spatial surfaces—*manifolds* is the technical term—with six curled-up dimensions. (These curled-up spaces are called Calabi-Yau manifolds.) In fact there are more than 7,000 ways (7,555 according to one calculation) to curl up six dimensions into one class of Calabi-Yau manifolds, and there are other classes. The different manifolds represent the way the dimensions curl up into different shapes.

The technical term describing this kind of the "shape of space" is *topology*, which refers to the ways that the points in the space are connected or related to each other. The simplest example to illustrate this is a cup and saucer, or maybe a ball and doughnut. A doughnut and coffee cup have the same topology, because by pulling and stretching you can deform one into the other. This means that points of the manifold are related in the same way (same topology). But you can't make a ball into a doughnut without cutting a hole. This changes the connections of the points—in other words, it changes the topology.

So of these thousands of possible topologies with six curled-up dimensions, which one is the topology of the real space we live in? It didn't seem possible to sort through all of them to check. But in 1995 one of those major breakthroughs came along that made the problem easier. Andy Strominger, who was then at the University of California, Santa Barbara (with collaborators David Morrison and Brian

Greene) found some new math showing that space seems to be constructed so that all the possibilities exist. Space of one shape can transform itself smoothly into any of the other shapes. In other words, we don't know how to transform a football into a doughnut, but nature does.

The most intriguing aspect of this advance was that once again the black hole entered the picture as the key link between gravity and quantum theory. It turns out that at the most fundamental of levels, described by quantum theory, a black hole is just like a basic particle of matter. Certain kinds of black holes can literally "condense" into a superstring particle.

At first glance, of course, black holes and superstrings seem quite different. So how can one turn into the other? The same way that water can turn into ice. Black holes and superstrings, like water and ice, are inherently the same thing. The technical term for one thing changing into its other self is *phase transition*.

When black holes become superstrings, the shape of space can transform from one possibility into a different one. In other words, there's a phase transition from one topology to another. Asking which space is the space we live in is like asking whether H_2O is a solid, liquid, or gas. It can be any of those. So in superstring theory, all those thousands of space-topology possibilities are just different versions of *one* space. And if there's only one way to represent space, that *must* be the space the universe is made of.

This realization didn't solve all the superstring problems, though. Another nagging issue was the existence of five different versions of superstring theory. In some versions, the strings were all closed loops; in others, some of the strings were just snippets with unattached ends. The different forms of superstring theory didn't all agree on how many dimensions space has, either. This was rather perplexing. If superstring theory was supposed to describe the whole universe, why would there need to be more than one version? "We seemed to have five distinct theories," John Schwarz said. "We really didn't understand why there should be five theories when we only need one—if it's the right one."[6]

Despite the advances during 1995, that problem remained. Nevertheless, it seemed to me that a lot of exciting science was going on at the frontiers of spacetime. But I hadn't seen anything yet. In 1996 a new tidal wave of research crashed onto the spacetime scene. It centered on a mysterious new development called M-theory.

Magic and Mystery

One of my favorite pastimes is to sit down at my computer every weeknight at 11 P.M. Central time and log on to *xxx.lanl.gov,* the best World Wide Web page in the universe. At that time, as if by magic, dozens of new papers by the world's leading (and following) physicists appear, grouped into a dozen categories. Anyone can browse the abstracts of these papers and then download and read or print out the full text. It is science in real time, the literally up-to-the-minute results of the latest in physics research.

During 1996 I became gradually aware that a new trend was showing up in theoretical physics, as the phrases "p-brane" and "D-brane" began appearing in more and more paper titles. I had no idea what a p-brane was, of course (other than a common expression that some scientists use to describe journalists). At one point I actually read a paper by John Schwarz that explained p-branes, but the full significance did not sink in.

But then in September I once again visited Princeton, where Frank Wilczek was to deliver an extended lecture at the Institute for Advanced Study on the frontiers of physics. Toward the end he mentioned the excitement about D-branes (which, as it turns out, are a special kind of p-brane). And while I was at the institute one of its administrators, Rachel Gray, gave me a copy of an article just out in *Nature* by another institute physicist, Edward Witten. In it he introduced the world to M-theory. The M, he said, could stand for magic, mystery, marvel, or membrane, according to taste.[7]

M-theory marked no less than a second superstring revolution, as big as the first one in the mid-1980s. This time the universe's basic units of existence were not loops of string behaving like tiny rubber bands, but something more like multidimensional soap bubbles known as supermembranes, thus the suffix *brane.*

The idea of a supermembrane was familiar to me. Back in 1988,

I'd written a column about the membrane idea as an offshoot of superstring theory. It hadn't seemed to have gone anywhere in the meantime. But I remembered well the physicist who had been pushing the membrane idea, Michael Duff of Texas A&M, a mere four-hour drive (actually three, but if I revealed that, I could be arrested for speeding) from Dallas.

So I arranged as soon as I could to drive to A&M to talk to Duff,[8] who brought me up-to-date on what was going on with membranes.

"We're all very excited by these new developments," he told me. "We're confident that we're on the right track." But he admitted that nobody really knew exactly where that track was headed. "If you try to pin me down about what M-theory is, I can't tell you," he said. "We've got glimpses of corners of M-theory, but the big picture is what we're still hoping to uncover."[9]

Superstring fans cheered the arrival of M-theory because it seemed to solve the problem of too many theories. There were five versions of string theory, but only one universe. Figuring out whether any of the five theories was the "right" one for the universe had been difficult. The math is so complex that efforts to solve the superstring equations required many approximations, an approach known in physics as perturbation theory. Using the approximation approach, string theory advanced slowly for years.

But then came a plot twist worthy of an Agatha Christie mystery, revealing the five theories to be one character wearing different disguises. M-theory, by avoiding the need for approximations, revealed unsuspected links connecting the five theories. "We thought we had five different theories, but they turned out to be locked together," said Joseph Polchinski of the University of California, Santa Barbara. "They turn out to be the same theory."[10]

M-theory's magic requires an extra dimension of space, for a total of eleven dimensions instead of the ten in superstring theory. And it added a dimension to its basic objects. Strings have just one dimension, length. Supermembranes have two dimensions, like the surface of a sphere, and therefore Duff said you could think of them as sort of like bubbles. To sum it all up, in the "old" superstring theory, the fundamental objects were one-dimensional strings living in ten-dimensional spacetime. In M-theory, the fundamental objects are two-dimensional membranes living in eleven-dimensional spacetime.

Of course, it's not that simple. Membranelike objects with even more dimensions can also exist. They are called p-branes (the objects that tipped me off that something was going on to begin with). The p in p-brane stands for the number of its dimensions. The eleven-dimensional spacetime of M-theory can contain objects with five dimensions, imaginatively designated five-branes, along with the two-dimensional membranes, or two-branes. When M-theorists talk about branes, they are not referring to how smart they all are.

By far the most famous of the branes in M-theory is the D-brane, the special p-brane that Wilczek described in his talk. (The D stands for Dirichlet, a nineteenth-century mathematician whose math is useful in their description.) In some versions of superstring theory, remember, the one-dimensional strings are not closed loops but open-ended snippets. D-branes provide surfaces, or something like "edges" in spacetime, for the open strings to end on. In this role, D-branes help to show how the five versions of superstring theory relate to one another.

The excitement over D-branes stems from their ability to aid in the understanding of black holes. In fact, a D-brane itself is like a mini–black hole, Strominger explained to me. "You can think of a black hole as built up of many tiny microscopic black holes," he said.[11] Those microscopic black holes are actually D-branes.

It turns out that D-branes showed a way to explain why black holes have a temperature. Hawking's discovery that black holes give off radiation allowed black holes to have a temperature, and therefore an entropy, as Bekenstein had surmised. (See chapter 10.) But the black hole's temperature was still a little mysterious. The whole idea of entropy suggested that a black hole's internal parts could be arranged in a number of different ways. Entropy is a measure of disorganization, after all, so something has to be disorganized. In a technical sense, the entropy of a system depends on how many different ways it can be organized, or the number of "microstates" that would present the same "big picture." (For example, the temperature of a gas is a measure of the average velocity of all the gas molecules. The individual molecules could have a whole variety of different speeds that would end up producing the same average, however.) With black holes, Hawking's formula related the black hole's temperature (and hence entropy) to its surface area, but there was no idea for what arrangement of "microstates" corresponded to that entropy.

But then in a breakthrough paper, Strominger and his colleague Cumrun Vafa from Harvard showed that a black hole's temperature depends on the number of ways the D-branes—those microscopic black holes—can be assembled. The Strominger-Vafa findings produced the same black-hole temperature results as those calculated by Hawking using entirely different methods. This was a striking result, suggesting that the D-brane approach had revealed something fundamental about black holes and the nature of spacetime. It turned out that spacetime seemed to be getting stranger and stranger.

After I'd visited Duff, I went to Pasadena to discuss M-theory with John Schwarz at Caltech, who was deep into D-branes. "The detailed mathematical structures of these D-branes seem to be telling us that our usual concepts of spacetime geometry break down at small distances," Schwarz said. "The usual idea of a nice smooth surface . . . is replaced by something much more bizarre."[12]

In fact, the very idea of distance fails to describe space on the scale of D-branes. The idea of an object's position becomes fuzzy, and repeated measurements of the distance between two stationary objects would give different answers, an odd state of affairs described by mathematicians as noncommutative geometry. "The spacetime itself is doing something weird," said Schwarz. "It's an idea which at least in the string theory context is very recent, and we're still kind of groping with. I think it will take some time to get it properly sorted out."[13]

Since then the sorting has produced some new wrinkles. Lately the hottest topic has been the idea that the extra dimensions of space in superstring theory aren't all as small as everybody thought.

One or more of those extra dimensions, some theorists now believe, could be as big as a millimeter across. If so, the visible universe could be one of many "parallel universes" crammed into this unseen space. While the universe we see looks big in three dimensions, it could be tiny in other dimensions—the way a sheet of paper can be long and wide but extremely thin. So many "big" universes could sit within millimeter-sized dimensions just the way many sheets of paper can fit into a slim file folder.

In this picture people cannot travel from one parallel universe to another because matter and energy are confined to the standard three dimensions. Only gravity is permitted to transmit its influence through the extra dimensions from one parallel universe to another. Experiments now in progress are attempting to detect subtle variations in the strength of gravity at distances smaller than a millimeter. If such deviations are found, they would be regarded as strong evidence that the extra dimensions really do exist.

The extra-dimension story does not immediately say a lot about information, but another hot new idea does—holography. That sounds like an invitation to return to the starship *Enterprise*'s holodeck, where laser beams generate lifelike three-dimensional images that simulate reality. But to physicists the "holographic principle" is another example of the power of viewing reality through the lens of information. (The holographic principle was developed, in fact, by two of the characters from chapter 10 engaged in analyzing the loss of information in black holes—Gerard 't Hooft and Leonard Susskind. Susskind's "black hole complementarity" discussed in chapter 10 was a precursor to the holography idea.)

It's hard to explain, because the physicists themselves don't fully understand it. But it involves a black hole's capacity to store information. The number of bits a black hole swallows is equal to the number of bits represented by its surface area—and a surface is two-dimensional. Somehow, all the information contained in three-dimensional space can be represented by the information available on that space's two-dimensional boundary. This is reminiscent of an ordinary hologram, where the appearance of three-dimensions is created from two-dimensional information. (In other words, you can have a flat picture, say on your credit card, that looks different if you view it from different angles.)

But ordinary holograms are just high-tech visual illusions, created by clever manipulation of light. With regard to black holes, the holographic principle contends that real three-dimensional information is being encoded on a two-dimensional surface. Translated into ordinary life, it would be like writing all the information contained in a book on its cover.

Within the context of string theory, this notion of holography seems to help explain some of the mysteries regarding black holes and information. But its deeper meaning isn't so clear. If the holographic principle is true in general, it would mean that somehow all the information in the 3-D world we live in is generated from 2-D information (in other words, the universe is a holodeck). Or in more mundane terms, whatever happens in a room could be recorded or somehow retrieved just by looking at what happens near the walls, says Nathan Seiberg.

But nobody knows if that's right. As it has been worked out so far, the holography idea applies just to space around black holes, and that space is very distorted compared to ordinary space. How the holography idea applies to ordinary "flat" space is now a subject of vigorous investigation.

"One of the main open questions is how to make all these holography ideas precise in flatland," Seiberg says. "It's high on the agenda of many people, but it looks like a very difficult question. Solving it will be a major breakthrough."[14]

Not Out of the Loop

Meanwhile, the loop quantum gravity advocates haven't been sitting on the sidelines. Twice in 1998 I attended meetings where Smolin's collaborator Carlo Rovelli spoke on the progress the loop approach had been making on some of the same problems.

The loop approach, Rovelli points out, offers a conceptual advantage over superstrings in regard to the origin of spacetime itself. Superstring theory assumes a "background" spacetime already sitting there, and the strings do their thing in that spacetime. The loop approach, on the other hand, constructs spacetime from scratch, piecing it together in the form of a spin network.

As for M-theory and its success in calculating the black hole entropy, Rovelli points out that the right answer can be computed only for certain special kinds of black holes. The loop approach gives approximately the right answer for black holes in general.[15]

All in all, it's clear that a lot more work needs to be done to sort out what's going on with black holes and what approach will turn out to give the deepest insight into the nature of spacetime. But it's pretty clear that the ideas of entropy and information will have an essential role to play in whatever understanding eventually emerges. And it's pretty significant, I think, that the details of the loop approach to black hole entropy touch on the intrinsic connection between information and existence that Wheeler has intuited. As Ashtekar and colleagues wrote in a 1998 paper, the spin network description of black hole spacetime turns up specific features suggesting an informationlike interpretation. "There is a curious similarity," wrote Ashtekar and his collaborators, "between our detailed results and John Wheeler's 'it from bit' picture of the origin of black hole entropy."[16]

Chapter 12

The Bit and the Pendulum

Science would perish without a supporting transcendental faith in truth and reality, and without the continuous interplay between its facts and constructions on the one hand and the imagery of ideas on the other. . . .

The possibility must not be rejected that several different constructions might be suitable to explain our perceptions.

—HERMANN WEYL,
Philosophy of Mathematics and Natural Science

Truman Burbank didn't know it, but he lived inside a gigantic dome. His life was a continuous TV show, his friends and families actors, his daily activities scripted, his future manipulated to maximize ratings, all without his knowledge. It made an entertaining movie—*The Truman Show* (Jim Carrey, Ed Harris, Paramount Pictures, 1998).

It was a film about perceptions, and about how reality and illusion merge into an indistinguishable mix. It was a story about the thin line separating fantasy from fact—if there is a line at all. But surely if there is a line, *The Truman Show* crossed it. Truman was born with the cameras rolling and for thirty years never ventured off the TV show's fictional island. How could anyone be deceived for so long into thinking that a stage set was real life? In the film, the show's mastermind director Christof (played by Ed Harris) was asked just that question.

"Why do you think that Truman has never come close to discovering the true nature of his world?" asked the interviewer.

"We accept the reality of the world with which we're presented," Christof replied.

And so do we all. Including the scientists who devote their lives to figuring out what reality is all about. Scientists are like the condemned man in Edgar Allan Poe's "The Pit and the Pendulum," groping in the dark to deduce the nature of his dungeon prison. Light seeping in radically reshaped the prisoner's perception of the dungeon, just as new ways of seeing nature recast the scientist's conception of reality. Such shifts in how reality appears inevitably raise philosophical questions about whether there is actually such a thing as "reality" out there, or whether we make it all up in our heads. That's an argument for another book. But there's a narrower issue about science itself that sometimes gets overlooked in these discussions, namely, that whatever the nature of what we call reality, science is not synonymous with it. Science is a description of reality.[1] Scientists who forget that distinction trap themselves in their paradigms, sort of the way Truman was trapped in his artificial-world TV show.

I think the clearest statement I have seen on this issue came from Hugh Everett, the creator of the many-worlds interpretation of quantum mechanics. In a short appendix accompanying a paper on his theory, he outlined the role of any scientific theory and its relationship to reality in a remarkably cogent way.

Every theory, Everett pointed out, has two parts. One is the "formal" part, the logical-mathematical structure, expressed as symbols and rules for manipulating them. The second part of a theory is its "interpretive" part, the rules for connecting the symbols to things in the real world—or perhaps more precisely, the "perceived world."[2] Presumably, in a good theory there is a strong link between the symbols of the theory and the perceptions of experience that those symbols represent. Nevertheless, the theory and its symbols remain a "model" of reality.

Often it's pretty clear that the theory is just a model that doesn't tell the whole story. Everett cites models of the atomic nucleus and "game theory" in economics as examples of theories in which the mathematical symbols clearly do not capture the totality of the phenomena they describe.

Yet sometimes, when science works extremely well, scientists begin to blur the distinction between their theory's models and reality. "When a theory is highly successful and becomes firmly established, the model tends to become identified with 'reality' itself, and the model nature of the theory becomes obscured," Everett wrote. This, he said, is precisely what happened with classical Newtonian physics. Newton's science was so successful that by the nineteenth century its validity was seldom questioned. Yet the twentieth century showed classical Newtonian science to be just a very good approximate model of reality. "The constructs of classical physics are just as much fictions of our own minds as those of any other theory," Everett declared. "We simply have a great deal more confidence in them."[3]

Everett concluded that there can be no such thing as the "correct theory." "There is nothing that prevents any number of quite distinct models from being in correspondence with experience (i.e., all 'correct')," he wrote.[4] In other words, science is not the same thing as the reality it describes. There is always a gap between reality and the description of it.

This point is similar to that made by the semanticist S. I. Hayakawa in his famous book *Language in Thought and Action*. The map is not the territory it depicts, Hayakawa stressed, just as in language, the word is not the thing it stands for. All sorts of misunderstandings ensue when the word is mistaken for the thing or the map is mistaken for the territory itself. Similar confusion arises when the current state of science is identified with the "true" nature of reality.

Nevertheless the merging of Newtonian physics with reality itself isn't that hard to understand, because it was rooted in the contemporary cultural understanding of what reality was like. The clockwork nature of the universe seemed evident to the deep thinkers of the centuries preceding Newton. Mechanical clocks kept time in harmony with the heavens. Newton showed how to make that conceptual understanding mathematically precise.

This new precision was embodied during Newton's time by a great improvement in mechanical clocks. Galileo set the stage for that development, just as he had paved the way to Newton's mathematics. Watching a lamp swinging in a cathedral, Galileo noticed the regularity of its motion. He figured out that the time it took to swing back and forth depended solely on the length of the swinging rod. (If it swung farther, it swung faster, so as always to take the same time to go from one end of its arc to another.) And Galileo showed that this rule applied to any such swinging rod, or pendulum.

Galileo saw that this insight provided the basis for building a better clock. Late in his life he sketched out a design. Apparently, though, a working model was never built.[5] Credit for inventing the pendulum clock fell to a contemporary of Newton's, the Dutch physicist Christiaan Huygens.

The pendulum clock provided science with a tool of unprecedented accuracy. Measuring time precisely was at the heart of Newton's laws. All sorts of other relationships in nature could be discerned through use of such a fabulous tool. But just as its predecessor mechanical clock became a metaphor, or symbol, as well as a tool, the pendulum clock took on greater significance in science. The pendulum itself became the standard model for explaining a vast range of physical systems. Textbooks even today use the pendulum as the prime example of a "harmonic oscillator," and physical phenomena are routinely analyzed by interpreting them in terms of harmonic oscillators in action.[6]

The pendulum's importance is only fitting, of course—it represented the technological perfection of the mechanical clock concept that had inspired Newtonian physics to begin with. But here again arises the need to distinguish between the word and the thing. For purposes of mathematical description, much of the world can be fruitfully viewed as pendulums in motion. But all the world is not a pendulum.

In the twentieth century the computer provided science with a new favorite tool. And it has naturally, like the pendulum centuries before, become science's favorite model for various other phenomena. Like Truman, and Poe's prisoner, science has discovered a new reality. The universe is not a clock after all. Nor is it a steam engine. We see now that it's a computer.

Of course, all the world is not really a computer, no more than it is a clock or a steam engine. These machines are metaphors for reality, not reality itself. But as the philosopher of science Stephen Toulmin has pointed out, such metaphors are powerful. "Calling the idea of the Machine a metaphor does something less than justice to its intellectual force," he writes. "We are all familiar with the risks that come from construing all scientific models in robustly realistic terms from the word go. Yet in actual fact, as time goes on, the concepts of science often acquire a richness and solidity that they initially lacked. A model begins as an analogy . . . but it later becomes more realistic and moves closer to a strictly 'literal' mode of speech."[7]

Thus viewing the world as a computer—analyzing natural phenomena in terms of information processing—can provide a new framework for explaining how the world works, just as the clock and steam engine have in the past. They were tools of enormous power, and they provided new ways of thinking about reality. With the mechanical clock, the universe became a deterministic mechanism. After the steam engine, the universe was viewed as a huge heat engine, running out of energy.

The computer is also a powerful tool, and has also provided a new way of thinking about how the world works. Everywhere around them, scientists see a world governed by the processing of information. Information is more than a metaphor—it is a new reality.

This information-computation approach has already opened up new avenues for exploring the mysteries of quantum physics, cells, brains, and black holes. It may in fact be in the process of restructuring all of science into a novel framework.

Critics of this point of view will surely argue that all the hype over information is unjustified and merely faddish. Such objections are not entirely without merit. In the 1950s, shortly after the electronic computer first appeared, many scientists promoted a computer-information–based approach to science; those efforts fizzled. "Cybernetics" was the buzzword back then, a "complex of ideas," wrote its inventor, Norbert Wiener, encompassing communication (information) theory, its relationship to the study of languages, computers, psychology, the nervous system, "and a tentative new theory of scientific method."[8] As the historian Peter Galison noted, cybernetics became a central feature in ongoing efforts by a number of scholars to frame a "unified science." "With enormous, perhaps overwrought enthusiasm, physiologists, sociologists, anthropologists, computer designers and philosophers leapt on the cyber bandwagon."[9] The new interdisciplinary fever afflicted even many physicists, Galison noted.

About the only aspect of cybernetics alive today is the "cyber" in cyberspace, the original discipline largely dismissed as another failed fad. So it's fair to ask why today's efforts to portray the world with an information-computer approach to science are any more significant. My answer is simply that the original enthusiasm over cybernetics was premature. It was not a point of view that could be imposed on science; it had to grow from within science.

After all, it took a long time for the mechanical clock to imprint its cultural influence into scientific thought, and steam engines of one sort or another had been around for a century before Carnot got around to inventing thermodynamics. It shouldn't be surprising that it has taken the computer a few decades to recondition the way that people think. The computer could provide the impetus for a truly new and deep understanding of nature only after it had infiltrated the way scientists think about what they do.

But now that is happening. In field after field, scientists explore their subject from a viewpoint colored by the computer, both as tool and as symbol. From biologists untangling the inner action of cells and neuroscientists analyzing the brain to physicists teleporting photons and building quantum computers, the idea of information as a tangible commodity sits at the core of their enterprise.

What does it all mean for science as a whole? Who knows? But to anyone curious about that question, it seems to me that two lines of thought provide the richest source of speculation: Wheeler's and Landauer's. Wheeler's slogan of "It from Bit" serves as a powerful inspiration for those who would seek deep meaning in the universe and an ultimate understanding of existence. Landauer's principle in turn provides the substance on which information science can build in the process of constructing such an understanding.

Landauer's Principle

Erasing information requires the dissipation of energy, Landauer's principle tells us. It sounds deceptively simple. But its implications are immense. It is the concrete example of the connection between physical reality and the idea of information as measured in bits. And in demonstrating that information is physical and real, it identifies the only true limit on the ability of computers to process information.

That limit, Landauer asserts, is not tied to the mistaken notion that computing inevitably requires energy. The limit is more cosmological in nature. And that limit threatens the very foundation of physics. In fact, Landauer has issued a warning of sorts to physicists at large—namely that the standard approach to physics today cannot in the long run be right. After all, he has often noted, what are the laws of physics if not recipes for performing computations? Scientists use laws to calculate what nature will do in advance of nature's doing it. Do such laws have any meaning if the calculations could not actually be performed? In other words, information processing and the laws of physics must in principle be tightly entangled. The ultimate laws must be limited to what actually can, in principle, be calculated in the universe we inhabit.

If calculating the outcome of any given law requires a computer with more memory than that available in the entire universe, then you might as well wipe that law off the books—it is worthless and meaningless.

This attitude defies the lessons that mathematicians have taught physicists, Landauer said in 1998 at a seminar at Ohio State University in Columbus.

"We've all grown up in the tutelage of the mathematician," Landauer said. "The mathematician has told us that we know what pi is, because we can calculate as many places of pi as we want. . . . Well, if I don't have a big enough memory, can I keep going on forever and ever? And in a universe which is very likely finite, can I really have a memory as big as I want?"[10]

And is it really possible in principle to make every step in the process absolutely 100 percent reliable with no chance of error? And what about all the many sources of degradation to the computing equipment—like cosmic rays, corrosion, and coffee spills? There's a question, Landauer maintained, "whether in the real physical world this process the mathematician has taught us about is implementable."

He has a strong suspicion that it's not.

"In other words, we're asking questions like how many degrees of freedom in the universe can be brought effectively together in principle, not within your NSF budget, but in principle, how many degrees can be brought effectively together in a computer? . . . I don't know the answer, but it's very likely to be limited and finite."

That is to say, a lot of conceivable calculations are not possible in the real world. The laws of physics we have now are idealizations. The "real" laws must be limited by the ability to crunch numbers in the universe as it is.

"Mathematics and computation, information handling, is done in the real physical world," Landauer observed. "Therefore it depends upon the laws of physics and the parts available in the physical universe."

But today's laws of physics come in the form of "continuum" mathematics. Newton's laws, the Schrödinger equation of quantum mechanics, Maxwell's equations describing electromagnetism—all are expressed in the calculus of continuous functions. Such equations imply the possibility of infinitely fine precision, requiring a computer that could store an infinite number of digits. "We don't know that these laws can be implemented in our physical universe," Landauer pointed out. "And sensible laws of physics must be implementable in our real physical universe, at least in principle. So we're looking for laws of physics which in fact are compatible with the information-handling capabilities that are really allowed in the universe. We're searching for a more self-consistent framework."

Landauer emphasized that he did not have the answers to all these mysteries. "I don't know this self-consistent formulation and I'm not the one who's going to invent it. Someone who knows far more about cosmology than I do will have to do that," he said. His speculations were really just a "want ad" for a better theory. But he saw more clearly than most physicists, I think, that our current understanding of nature is incomplete in a way different from what most other physicists ever contemplate. Somehow there must be a different framework for understanding the universe that hasn't revealed itself yet, and the success of the information-processing approach to science may offer a clue about where to look for that missing framework.

In his Columbus talk, Landauer offered two examples of where his view might lead to progress. One is in understanding the irreversible loss of useful energy in the universe, closely related, of course, to the second law of thermodynamics and the arrow of time. It has long been a great mystery why the laws of physics seem to allow time to flow either forward or backward, yet real life is governed by a one-way clock. Landauer thinks the physics of information may offer a way to understand that asymmetry.

To illustrate his idea, he chose—naturally—the pendulum. In real life, a common way of dissipating energy is friction, and the world is full of friction, Landauer pointed out. "The normal way of explaining that is to say, well, the system you're looking at is not the whole world; there's more to it than that. And there's stuff out here that can take energy away from here—that's friction." This is a useful point of view, but somewhat unsatisfactory, because it requires drawing an arbitrary boundary between the system and the world outside. What if you're trying to understand the whole universe, with nothing "outside" to siphon off energy?

So Landauer looked at the situation this way. The question is, if you put energy into a system, can you get it back out? If you can, then the situation is reversible, you have no dissipation of energy. And often that is pretty easy to do. Take the pendulum. Give it a kick, and it starts swinging—your foot has imparted energy to the pendulum. Now, can you get that energy back? Yes! Wait for half a swing, Landauer says, and give it another kick identical to the first one. Only this time the pendulum will be headed back toward your foot. If you kick at precisely the right time with the same energy, the pendulum will stop, and its energy will flow back into your foot. (Of course, if you want to use the energy again, you'd want to design a better energy receiver than a shoe.) To the extent that pendulums represent physical systems, then, you can always get the energy you put in back out, and the pendulum (or the system it represents) merely serves as an energy storage device. You can withdraw the energy whenever you like.

"But to do this," Landauer pointed out, "you have to understand something about the motion of the system. You'd have to know the pendulum law, you'd have to know when to apply the second kick."

And for a pendulum, that's all very simple. But in complicated systems, getting the energy back wouldn't be so easy. It would require a delicate calculation to determine precisely when to apply that second kick.

In some cases, for sufficiently complicated systems, it could be impossible, in principle, to make that calculation. If you cannot calculate when to apply the second kick, you can't get the energy all back—some of it is lost, dissipated. "Maybe that's the real source of dissipation," Landauer said.[11]

The second example Landauer raised for the usefulness of his viewpoint involved quantum mechanics. Perhaps the underlying reason for the universe's classical appearance—despite its true quantum mechanical nature—may bear some relationship to the limits on computation.

"We know the world is quantum mechanical at the bottom, and yet it looks very classical," Landauer said in Columbus. Why doesn't it look quantum mechanical? Well, quantum mechanics is all about waves, and what happens depends a lot on whether these waves are in phase or out of phase—in other words, how they interfere with one another. A fine-grained quantum description of reality therefore requires a lot of precise computation about phases, and in principle it makes no sense to say those phases could be calculated to an arbitrarily accurate precision. Therefore quantum mechanics acquires some intrinsic "coarse graining" in a universe where calculations cannot be infinitely precise, and the coarse graining leads to the classical appearances that people perceive.[12]

All this led Landauer to the conclusion that the very nature of physical law is radically different from the standard view. That view, he said, is ingrained so deeply into culture that it is even expressed in the Bible—"In the beginning was the word"—or, in the beginning, the controlling principle of the universe was present from the outset. To put it another way, the laws of physics preceded existence itself. But Landauer argued otherwise. Any "controlling principle" must be expressed within the physical features of the universe. "We're saying, hey, the controlling principle of the universe has to be expressed in physical degrees of freedom," Landauer said, "and it wasn't there in the beginning."

Law without Law

John Wheeler has long expressed a similar view. He expresses it with the slogan "law without law." "Are the laws of physics eternal and immutable?" Wheeler asks. "Or are these laws, like species, mutable and of 'higgledy-piggledy' origin?"[13] Wheeler votes for "higgledy-piggledy." He also votes, like Landauer, against the concept of a continuum of numbers. Standard continuum mathematics holds that between any two numbers, you can always find another number, ad infinitum. But the whole continuum concept is an abstraction, without physical basis.

"No continuum," Wheeler insists. "No continuum in mathematics and therefore no continuum in physics. A half-century of development in the sphere of mathematical logic has made it clear that there is no evidence supporting the belief in the existential character of the number continuum."[14]

Other concerns of Wheeler's are similar to Landauer's, though not identical. Wheeler has very much hoped for some breakthrough in understanding the origin—the reasons for—quantum mechanics. "How come the quantum?" he repeats again and again—most recently to me in a letter of December 15, 1998. "Miserable me," Wheeler wrote. "I still don't know 'How come the quantum?' and still less how come existence."[15]

The best idea he has been able to come up with to point toward an answer is "It from Bit." In his grandest paper on the subject, he argued the case with his usual eloquence.

"It from bit symbolizes the idea that every item of the physical world has at bottom—at a very deep bottom, in most instances—an immaterial source and explanation; that what we call reality arises in the last analysis from the posing of yes-no questions and the registering of equipment-evoked responses; in short, that things physical are information-theoretic in origin," Wheeler wrote.[16]

When we ask whether a photon detector clicked within a certain one-second interval, and the answer is yes, we say "a photon did it," Wheeler says. The implication is that a photon had been emitted by a source and traveled to the detector. There is clearly no sense in saying that the photon existed before the emission or after the detection. "However," Wheeler points out, "we also have to recognize that any talk of the photon 'existing' during the intermediate period is only a blown-up version of the raw fact, a count."[17] The count tells us the answer to the initial question, Did the counter click? We can say it's because of a photon, but in the end all we really know is the raw fact, the answer to a yes-no question. A bit.

Experiments involving electromagnetic fields can be analyzed in a similar way. "Field strength . . . reveals itself through shift of interference fringes, fringes that stand for nothing but a statistical pattern of yes-or-no registrations," Wheeler notes.[18]

And then, of course, the surface area of a black hole reflects the number of bits it contains—that is, "the number of bits, that would be required to specify in all detail the configuration of the constituents out of which the black hole was put together."[19] These are examples, but the black hole case in particular points to the intimate relation between information and quantum physics—the heart and soul of physics.[20] "Tomorrow," Wheeler predicts, "we will have learned to understand and express all of physics in the language of information."[21]

Quantum information theory is clearly a step in that direction. The reality of quantum teleportation, quantum cryptography, and quantum computers has begun to force upon physics the recognition that information is a basic aspect of existence. Still, I couldn't promise that "It from Bit" will be the answer to all the universe's riddles. It remains outside the mainstream of physics today, and many other fruitful routes of investigation are being traveled. Some appear to be paths of great promise toward deeper understanding of everything. Investigations into the topology of space, noncommutative geometry, D-branes, M-theory, and/or loop quantum gravity may provide the self-consistent and complete formulation of physics that will satisfy everybody someday. On the other hand, it is just possible that putting these pieces together into that consistent picture will be facilitated by the appreciation of information.

In any case, it is inescapable that Wheeler's image of the bit-covered black hole is where many of these issues come together.

"At present, the greatest insights into the physical nature of quantum phenomena in strong gravitational fields comes from the analysis of thermodynamic properties associated with black holes," the physicist Robert Wald of the University of Chicago has written. He thinks that pursuing those insights will lead to radical revisions of standard ideas about reality. "I believe that the proper description of quantum phenomena in strong gravitational fields will necessitate revolutionary conceptual changes in our view of the physical world," Wald writes.[22]

I think that whatever new way of construing reality emerges, it will turn out—in some fashion or another—to involve bits. If so, the imagery of the computer will fully join that of the clock and steam engine as ways of representing reality. That will not make reality any less real or more arbitrary. It will merely reflect what Hermann Weyl astutely noted when he said that science depends both on "faith in truth and reality" and on the interplay of facts with imagery. In science, image and truth are inevitably intertwined.

And that makes the image of a black hole as bottomless pit particularly apt. I realized that just the other day watching the movie *A Civil Action,* in which the lawyer played by Robert Duvall offers John Travolta some friendly advice: "If you're really looking for the truth," Duvall declares, "look for it where it is—at the bottom of a bottomless pit."

Notes

Introduction

1. Steven Weinberg, interview in Austin, Tex., Oct. 19, 1989.
2. John Wheeler, "Information, Physics, Quantum: The Search for Links," in Anthony J. G. Hey, ed., *Feynman and Computation* (Reading, Mass.: Perseus Books, 1999), 309. Reproduced from the Proceedings of the Third International Symposium on the Foundations of Quantum Mechanics, Tokyo, 1989, 354–68.
3. John Wheeler, *Geons, Black Holes, and Quantum Foam* (New York: W. W. Norton, 1998), 240.
4. Rolf Landauer, interview, Nov. 13, 1990.
5. Shmuel Sambursky, ed., *Physical Thought from the Presocratics to the Quantum Physicists* (New York: Pica Press, 1975), 47.
6. Werner Heisenberg, *Physics and Philosophy* (New York: Harper & Row, 1958), 63.
7. Wheeler, "Information, Physics, Quantum," 322.

Chapter 1: Beam Up the Goulash

1. Robert Park, "What's New," Feb. 2, 1996, *www.aps.org/WN/WN96/wn020296 .html.*
2. Samuel Braunstein, "A fun talk on teleportation," Feb. 5, 1995. Available at *http://www.research.ibm.com/quantuminfo/teleportation/braunstein.html.*
3. Richard Feynman, *The Character of Physical Law* (Cambridge, Mass.: MIT Press, 1967), 129.
4. B. L. van der Waerden, ed., *Sources of Quantum Mechanics* (New York: Dover, 1967), 25.
5. Niels Bohr, *Essays 1958–1962 on Atomic Physics and Human Knowledge* (New York: Vintage Books, 1966), 76.
6. Charles Bennett, interview at IBM Thomas J. Watson Research Center, Nov. 12, 1990.
7. Charles Bennett, news conference, American Physical Society meeting in Seattle, Wash., Mar. 24, 1993.
8. Benjamin Schumacher, interview in Santa Fe, N.M., May 15, 1994.
9. Charles Bennett, interview in Santa Fe, N.M., May 15, 1994.
10. Seth Lloyd, interview in Santa Fe, N.M., May 17, 1994.

Chapter 2: Machines and Metaphors

1. The basic idea of the Turing test is to pose questions to an unseen answerer, either a human or a computer. If the answers come from a computer, but the questioner can't deduce whether it's a computer or a human, the computer passes the test and supposedly should be regarded as intelligent.

2. I have no way of knowing, of course, whether Oresme encountered clocks as a teenager or only much later. One writer speculates that the great clock at the palace of France's King Charles V—built in 1362—may have served as Oresme's inspiration.

3. Nicole Oresme, *Le livre du ciel et du monde* (Madison: University of Wisconsin Press, 1968), 283.

4. Ibid.

5. Sadi Carnot, *Reflections on the Motive Power of Fire*, trans. Robert Fox (New York: Lilian Barber Press, 1986), 63.

6. Joel Shurkin, *Engines of the Mind* (New York: W. W. Norton, 1984), 298.

7. I'm neglecting the abacus, which some people would consider a calculating device, because somehow to me it doesn't seem to be very mechanical. And I could never understand how it works. Also, it seems that a German named Wilhelm Shickard designed a calculating machine even better than Pascal's in 1623—the year Pascal was born. But Shickard's machine was destroyed by fire while still under construction, and Shickard died in 1635. His design was rediscovered more than three centuries later, so his work played no part in the history of computing as it developed.

8. E. T. Bell, *Men of Mathematics* (New York: Simon & Schuster, 1937), 73.

9. Martin Campbell-Kelly, ed., *Charles Babbage: Passages from the Life of a Philosopher* (New Brunswick, N.J.: Rutgers University Press, 1994), 13.

10. Ibid., 27.

11. Quoted in Herman Goldstine, *The Computer: From Pascal to von Neumann* (Princeton, N.J.: Princeton University Press, 1993), 35.

12. George Boole, *An Investigation of the Laws of Thought* (1854; reprint, New York: Dover, 1958), 69.

13. Andrew Hodges, *Alan Turing: The Enigma* (New York: Simon & Schuster, 1983), 97.

14. The full title was "On Computable Numbers with an Application to the *Entscheidungsproblem*," the German name for the decidability problem. It was published in the *Proceedings of the London Mathematical Society* 42 (1936): 230–65.

15. This is just my way of capturing the essence of many different expressions of this thesis. For example, John Casti (*New Scientist*, May 17, 1997, 31) defines it this way: "A Turing machine running a suitable program can carry out any process that can reasonably be called a computation." M. A. Nielsen (*Physical Review Letters*, Oct. 13, 1997, 2915) expresses it as the thesis that the class of "computable functions . . . corresponds precisely to

the class of functions which may be computed via what humans intuitively call an algorithm or procedure." Wojciech Zurek's version (in *Complexity, Entropy and the Physics of Information*, 81) is "What is human computable is universal computer computable."

Chapter 3: Information Is Physical

1. Edward Fredkin, interview in Dallas, Oct. 3, 1992.
2. Rolf Landauer, interview at IBM Thomas J. Watson Research Center, Apr. 16, 1997.
3. Wojciech Zurek, interview in Brighton, England, Dec. 19, 1990.
4. Arthur Eddington, *The Nature of the Physical World* (Ann Arbor, Mich.: University of Michigan Press, 1958), 74.
5. Harvey S. Leff and Andrew F. Rex, eds., *Maxwell's Demon* (Princeton, N.J.: Princeton University Press, 1990), 4.
6. Szilard also analyzed the case of a single molecule, to show that the Second Law was not merely a statistical consequence of dealing with large numbers.
7. Wheeler pointed out once that dictionaries don't list all the words in the language, so the English language probably consists of about a million words. Specifying one word out of a million would represent 20 bits. Since one bit is the equivalent of answering a single yes-no question, picking a word out of the English language therefore corresponds to playing the party game of Twenty Questions. You can win just by asking first if the word is in the first half of the dictionary, and then narrowing it by half again with each question.
8. Basically, the number of bits is equal to the logarithm of the number of choices, using the binary (base 2) system of numbers.
9. Tom Siegfried, "Exploring the relationship of information to existence," *Dallas Morning News*, May 14, 1990, 7D.
10. Jeremy Campbell, *Grammatical Man* (New York: Simon & Schuster, 1982), 52.
11. Wojciech Zurek, *Complexity, Entropy and the Physics of Information* (Redwood City, Calif.: Addison-Wesley, 1990), vii.
12. Rolf Landauer, "Computation: A Fundamental Physical View," *Physica Scripta* 35 (1987), reprinted in Harvey S. Leff and Andrew F. Rex, eds., *Maxwell's Demon* (Princeton, N.J.: Princeton University Press, 1990), 260.
13. Rolf Landauer, interview at IBM Thomas J. Watson Research Center, May 16, 1995.
14. Ibid.
15. Donald Cardwell, *The Norton History of Technology* (New York: W. W. Norton, 1995), 113–16.
16. Parent did not investigate "overshot" wheels in which the water fills containers at the top of the wheel, thus getting an assist from gravity in turning the wheel around.

17. Cardwell, *Norton History of Technology*, 115–16.
18. Rolf Landauer, "Information Is Inevitably Physical," in A.J.G. Hey, ed., *Feynman and Computation* (Reading, Mass.: Perseus, 1999), 78.
19. Charles Bennett, news conference, American Physical Society meeting in Seattle, Wash., Mar. 24, 1993.
20. "When I first explained this to Feynman, he didn't believe it," Bennett told me. "After fifteen minutes or so, though, he realized it would work. Well, maybe not fifteen—maybe it was only five."
21. One was Fredkin, who devised a scheme for reversible computing involving imaginary billiard balls.
22. Ralph Merkle, American Physical Society meeting in Seattle, Wash., Mar. 24, 1993.

Chapter 4: The Quantum and the Computer

1. Hans Bethe, interview at Cornell University, Apr. 14, 1997.
2. Richard Feynman, "Simulating Physics with Computers," *International Journal of Theoretical Physics* 21 (1982), 486.
3. Ibid., 474.
4. Feynman's conjecture was later worked out in detail by Seth Lloyd, who presented a talk about it at the American Association for the Advancement of Science meeting in Baltimore in 1996.
5. It is not necessary to imagine multiple universes for a quantum computer to work. Deutsch thinks that's the most natural explanation, but Charles Bennett emphasizes that you expect the same quantum computing results regardless of your interpretation of quantum mechanics.
6. The details are technical. But if you're interested, it involves finding the period of a periodic function. When a function is periodic, the solution repeats itself at regular intervals, or periods. The multiple computations in a quantum computer can "interfere" with each other the way water waves either boost or cancel each other. In a quantum computer running Shor's algorithm, the various computations interfere in such a way that most of them are canceled out, and the computer spits out an answer for the period of the periodic function. With that number, finding the prime factors is easy. The problem of choosing the right answer out of trillions of possibilities is avoided in this case because there is no need to preserve all the information contained in all those quantum possibilities—the calculation is designed specifically so that all but one of the answers will be destroyed on purpose. The one that remains is the one you want.
7. Some of my science writing colleagues will scoff at this comparison, suggesting that the AAAS meeting is more like the NBA semifinals.
8. Isaac Chuang, interview in Palm Springs, Calif., Feb. 18, 1998.
9. John Preskill, interview in Palm Springs, Calif., Feb. 18, 1998.
10. Jeff Kimble, interview in Pasadena, Calif., Jan. 19, 1996.

Chapter 5: The Computational Cell

1. The Rivest-Shamir-Adelman, or RSA, system, the code that Peter Shor's algorithm can break with a quantum computer.
2. Leonard Adleman, interview in Los Angeles, Mar. 17, 1998.
3. Ibid.
4. Leonard Adleman, "Computing with DNA," *Scientific American* 279 (Aug. 1998): 61.
5. Actually, Crick was convinced almost instantly, but Watson was a little hesitant, Crick reports in his autobiographical book, *What Mad Pursuit*. But Watson came around soon thereafter.
6. Francis Crick, *What Mad Pursuit* (New York: Basic Books, 1988), 66.
7. Since an authority as illustrious as Crick had proclaimed this scenario a dogma, many scientists considered it rather like a holy pronouncement that must be invariably true—or that at least must not be doubted. But they had not realized what Crick thought of religion. In fact, he later recounted, he chose "dogma" because he was familiar with its use in religion. He thought a religious dogma was "a grand hypothesis that, however plausible, had little direct experimental support." "I thought that all religious beliefs were without any serious foundation," Crick wrote in his autobiography, *What Mad Pursuit* (page 109). As it turns out, there are important exceptions to the central dogma, although it does describe the ordinary flow of information from gene to protein.
8. Francis Crick, interview in La Jolla, Calif., June 10, 1998.
9. An allusion to *The Hitchhiker's Guide to the Galaxy*, a series of novels by Douglas Adams.
10. Laura Landweber, interview in Princeton, N.J., Oct. 21, 1998.
11. Ibid.
12. Ibid.
13. Dennis Bray, "Protein molecules as computational elements in living cells," *Nature* 376 (July 27, 1995): 307.
14. Ibid., 312.
15. Tom Siegfried, "Computing with chemicals," *Dallas Morning News*, Feb. 3, 1992, 9D.
16. Elliott Ross, "Twists and Turns on G-protein Signalling Pathways," *Current Biology* 2 (1992): 517–19.
17. Stuart Kauffman, *The Origins of Order* (New York: Oxford University Press, 1993); and *At Home in the Universe* (New York: Oxford University Press, 1995).
18. P. D. Kaplan, G. Cecchi, and A. Libchaber, "DNA based molecular computation: Template-template interactions in PCR," preprint at *asterion.rockefeller.edu/cgi-bin/prepform*, 1.
19. Leonard Adleman, "Molecular Computation of Solutions to Combinatorial Problems," *Science* 266 (Nov. 11, 1994): 1021–24.

20. Mitsunori Ogihara and Animesh Ray, "Simulating Boolean Circuits on a DNA Computer," preprint at *www.cs.rochester.edu/users/faculty/ogihara/research/pubs.html*, August 1996.
21. David Gifford, "On the Path to Computation with DNA," *Science* 266 (Nov. 11, 1994): 993.

Chapter 6: The Computational Brain

1. Herman Goldstine, *The Computer: From Pascal to von Neumann* (Princeton, N.J.: Princeton University Press, 1993), 182.
2. Ibid.
3. Norman Macrae, *John von Neumann* (New York: Pantheon Books, 1991), 9. I heard of this anecdote in a talk by Melanie Mitchell of the Santa Fe Institute.
4. John von Neumann, *The Computer and the Brain* (New Haven, Conn.: Yale University Press, 1958), 39.
5. Jeremy Bernstein, *The Analytical Engine* (New York: Random House, 1964), 24.
6. John Von Neumann, *The Computer and the Brain*, 43–44.
7. Christof Koch, *Biophysics of Computation* (New York: Oxford University Press, 1999), 1.
8. News conference, annual meeting of the Society for Neuroscience, Los Angeles, Nov. 8, 1998.
9. Ira Black, *Information in the Brain* (Cambridge, Mass.: MIT Press, 1991), 61.
10. News conference, annual meeting of the Society for Neuroscience, Anaheim, Calif., Oct. 26, 1992.
11. I will get in trouble here for leaving somebody out, but my favorite institutes (not including universities and institutes associated with them) include Salk, the Institute for Advanced Study in Princeton, the Santa Fe Institute in New Mexico, and Fermilab in Illinois (okay, it's a lab, not an institute). If your institute is not on this list of the top half dozen, it is one of the two I didn't mention.
12. News conference, annual meeting of the Society for Neuroscience, Anaheim, Calif., Oct. 26, 1992.
13. Ibid.
14. Ibid.
15. This is a rather sketchy picture of an intricate mathematical argument. You can get the gory details in chapter 4 of Penrose's book *The Emperor's New Mind* (New York: Oxford University Press, 1989).
16. John Searle, "Is the Brain's Mind a Computer Program?" *Scientific American* 262 (Jan. 1990): 26–31.
17. John Searle, interview in Richardson, Tex., Mar. 3, 1988.
18. Paul Churchland and Pat Churchland, "Could a Machine Think?" *Scientific American* 262 (Jan. 1990): 32–37.

19. Terrence Sejnowski, interview in Anaheim, Calif., Oct. 27, 1992.
20. Rolf Landauer, interview at IBM Watson Research Center, May 16, 1995.

Chapter 7: Consciousness and Complexity

1. Francis Crick, interview in La Jolla, Calif., June 10, 1998.
2. Ibid.
3. Ibid.
4. E. T. Bell, *Men of Mathematics* (New York: Simon & Schuster, 1937), 526.
5. It's a journal of real-time debate, with a target article taking a position that is then assaulted (or supported) by numerous commentaries published in the same issue. It beats waiting for a subsequent issue's letters to the editor.
6. Tim van Gelder, "The Dynamical Hypothesis in Cognitive Science," preprint at *www.arts.unimelb.edu/au/~tgelder/Publications.html*.
7. Melanie Mitchell, "A Complex-Systems Perspective on the 'Computation vs. Dynamics' Debate in Cognitive Science," Santa Fe Institute working paper 98-02-017 at *www.santafe.edu/sfi/publications/98wplist.html*, Feb. 1998, 2.
8. Jim Crutchfield, "Dynamical Embodiments of Computation in Cognitive Processes," Santa Fe Institute working paper 98-02-016 at *www.santafe.edu/sfi/publications/98wplist.html*, Feb. 1998.
9. Melanie Mitchell, interview in Santa Fe, Sept. 17, 1997.
10. They are called the Ulam lectures, given in memory of the mathematician Stanislaw Ulam.
11. This is a one-dimensional cellular automaton, just a line of lights, so each bulb has only two neighbors. You could of course imagine, or even build, a more sophisticated cellular automaton, with a two-dimensional grid. And you could also have a three-dimensional automaton with even more neighbors per bulb. (In a real human brain, one nerve cell is typically connected to thousands of others.) This gets much more complicated to visualize, and includes many more behavior possibilities for the system, but the principles are the same. It's just much easier to talk about a single string of Christmas tree lights.
12. William Ditto, personal e-mail communication, Sept. 22, 1998.
13. Ibid.
14. Francis Crick, interview in La Jolla, Calif., June 10, 1998.
15. There are a few scientists from other categories who do support him, however, including the physicist Dimitri Nanopoulos.
16. Part of Penrose's idea involves the belief that gravity is responsible for eliminating multiple quantum possibilities. I have heard of plans to test this idea experimentally, but such tests are probably still years away. In any case the physicist Max Tegmark has calculated that normal interaction with the environment would destroy the multiple quantum possibilities far faster than gravity would—and far faster than necessary for quantum effects to play a role in consciousness. See Max Tegmark, "The Quantum Brain," preprint quant-ph/9907009 at *xxx.lanl.gov*.

17. Francis Crick, interview in La Jolla, Calif., June 10, 1998.
18. Lawrence Weiskrantz, *Consciousness Lost and Found* (New York: Oxford University Press, 1997), 62–66.
19. Michael Gazzaniga, *Nature's Mind* (New York: Basic Books, 1992), 15.
20. Ibid., 59.
21. After one column about his work, I received a congratulatory letter from Edelman for representing his ideas so clearly and concisely. Apparently he found this to be a rare occurrence.
22. Gerald Edelman, *Bright Air, Brilliant Fire* (New York: Basic Books, 1992), 112.
23. I should point out that Francis Crick is not impressed. He calls the theory neural Edelmanism.
24. Annual meeting of the Cognitive Neuroscience Society, San Francisco, Apr. 7, 1998.
25. Giulio Tononi and Gerald Edelman, "Consciousness and Complexity," *Science* 282 (Dec. 4, 1998), 1849.

Chapter 8: IGUSes

1. John Wheeler, interview in Princeton, N.J., Feb. 10, 1998.
2. Freeman Dyson, *Disturbing the Universe* (New York: Basic Books, 1979), 250.
3. John Wheeler, "Hermann Weyl and the Unity of Knowledge," in *At Home in the Universe* (Woodbury, N.Y.: AIP Press, 1994), 185–86.
4. John Barrow, "Cosmology and the Origin of Life," preprint astro-ph/9811461 at *xxx.lanl.gov,* Nov. 30, 1998, 10.
5. John Barrow and Frank Tipler, *The Anthropic Cosmological Principle* (New York: Oxford University Press, 1986), 16–23.
6. Martin Gardner, *The New York Review of Books,* May 8, 1986, as reprinted in Gardner's *Whys & Wherefores* (Chicago: University of Chicago Press, 1989), 228.
7. Tom Siegfried, "Return to the center of the cosmos," *Dallas Morning News,* Feb. 9, 1987.
8. Ibid.
9. Heinz Pagels, American Association for the Advancement of Science annual meeting, Boston, February 1988.
10. In some current cosmological scenarios, the universe is a collection of multiple "inflated regions" or bubbles with different properties; we live in a bubble where the properties are hospitable to life, a new twist on the weak anthropic principle as an explanation for why we're here.
11. Computer users know that memory storage is typically measured in bytes, which are not the same as bits, but are made up of bits.
12. Murray Gell-Mann, American Association for the Advancement of Science annual meeting, Chicago, Feb. 11, 1992.
13. Ibid.
14. Ibid.

15. That would not, however, constitute a true "theory of everything." As Gell-Mann put it: "Now suppose we have the fundamental law for the elementary particles and for the universe. Then what? Is it a so-called theory of everything? Baloney! It couldn't possibly be a theory of everything because the theory is quantum mechanical. And the most quantum mechanics can do is to specify the relative probabilities of different histories of the universe."

16. Various definitions of depth and further measures of information and complexity are discussed by Gell-Mann and Seth Lloyd in "Information Measures, Effective Complexity, and Total Information," *Complexity* (Sept. 1996): 44–52.

17. Warren Weaver, "Recent Contributions to the Mathematical Theory of Communication," in Claude Shannon and Warren Weaver, *The Mathematical Theory of Communication* (Urbana, Ill.: University of Illinois Press, 1949), 8–9.

18. Gleick described Crutchfield as "short and powerfully built, a stylish windsurfer." I have to admit that it didn't occur to me to ask Crutchfield about windsurfing.

19. James Crutchfield, interview in Berkeley, Calif., Jan. 13, 1995.

20. Ibid.

21. Ibid.

22. Ibid.

23. Ibid.

24. James Crutchfield, "The Calculi of Emergence: Computation, Dynamics and Induction," *Physica D* 75 (1994), 11–54.

25. This is similar to Edelman's ideas (chapter 7) about "value centers" in the brain guiding behavior in response to input from the senses. Crutchfield uses different language to describe essentially the same idea.

26. One reason why I think John Horgan's book *The End of Science* is bunk.

27. James Crutchfield, interview in Berkeley, Calif., Jan. 13, 1995.

28. Murray Gell-Mann, American Association for the Advancement of Science annual meeting, Chicago, Feb. 11, 1992.

Chapter 9: Quantum Reality

1. Murray Gell-Mann, annual meeting of the American Association for the Advancement of Science, Chicago, Feb. 11, 1992.

2. Wojciech Zurek, interview in Los Alamos, N.M., Apr. 26, 1993.

3. John Wheeler makes this point in his book *A Journey into Gravity and Spacetime*. "Einstein emphasized to me that Newton understood the problems inherent in the concept of absolute space," Wheeler wrote. "Newton had courage, Einstein explained, in that he knew better than his followers that the concept of absolute space could not be defended; nonetheless, he realized that he had to assume the existence of absolute space in order to understand motion and gravity." I think Bohr in the same way realized the need to assume a classical world in order to understand quantum physics.

4. Everett, who went to work for the Pentagon rather than taking up a physics career, died of a heart attack in 1982.

5. Max Jammer, *The Philosophy of Quantum Mechanics* (New York: John Wiley & Sons, 1974), 517.

6. John Wheeler and Wojciech Zurek, eds. *Quantum Theory and Measurement* (Princeton, N.J.: Princeton University Press, 1983), 157.

7. See chapter 8. Einstein made that comment in John Wheeler's relativity seminar, Apr. 14, 1954.

8. Hugh Everett, "The Theory of the Universal Wave Function," in Bryce De-Witt and Neill Graham, eds., *The Many-Worlds Interpretation of Quantum Mechanics* (Princeton, N.J.: Princeton University Press, 1973), 117.

9. John Wheeler, "Law without Law," in *Frontiers of Time* (Austin, Tex.: Center for Theoretical Physics, University of Texas at Austin, 1978), 6, and telephone interview, Oct. 15, 1987.

10. Bryce DeWitt, "Quantum Mechanics and Reality," *Physics Today* 23 (Sept. 1970).

11. Deutsch, in fact, viewed quantum computing as evidence favoring the many-worlds interpretation, because you can imagine that a quantum computer gets its enhanced power by being able to compute in many universes at once. Charles Bennett disagrees. All major interpretations of quantum mechanics predict the same results for a quantum system that proceeds undisturbed and then is measured, and a quantum computer is just that sort of system. The success of a quantum computer, Bennett says, "is no more a rigorous proof of the many worlds interpretation than the ability of light to propagate in a vaccum is a proof of the existence of the luminiferous ether." (Bennett e-mail of June 15, 1994.)

12. Max Tegmark, "The Interpretation of Quantum Mechanics: Many Worlds or Many Words?" preprint quant-ph/9709032 at xxx.lanl.gov, Sept. 15, 1997.

13. Wojciech Zurek, interview in Los Alamos, N.M., Apr. 26, 1993.

14. Roland Omnès, "Consistent Interpretations of Quantum Mechanics," *Reviews of Modern Physics* 64 (Apr. 1992): 355.

15. Things get really technical here—the term *decoherence* is used in slightly different ways in different contexts. You should not attempt to define *decoherence* at home. For a brief discussion of this point, try Murray-Gell Mann and Jim Hartle's comments in J. J. Halliwell et al., *The Physical Origin of Time Asymmetry* (Cambridge, Mass.: Cambridge University Press, 1994), 341–43. Also see Murray Gell-Mann and James Hartle, "Quantum Mechanics in the Light of Quantum Cosmology," in Zurek, W. H., ed. *Complexity, Entropy and the Physics of Information* (Redwood City, Calif.: Addison-Wesley, 1990), 430.

16. Murray Gell-Mann, interview in Chicago, May 13, 1993. Zurek still defends his approach, arguing that it explains all observable phenomena. "In the context of wherever there is an observer that's able to pose questions, you always end up having to split the universe into subsystems," he said.

17. Murray Gell-Mann, annual meeting of the American Association for the Advancement of Science, Chicago, Feb. 11, 1992.

18. Originally, Gell-Mann and Hartle referred to quasiclassical "domains," but in later revisions of their work they switched to "realms" to avoid confusion with other uses of the word domain in physics.

19. Murray Gell-Mann, annual meeting of the American Association for the Advancement of Science, Chicago, Feb. 11, 1992.

20. Murray Gell-Mann, interview in Chicago, May 13, 1993.

21. Murray Gell-Mann and James Hartle, "Equivalent Sets of Histories and Multiple Quasiclassical Domains," preprint gr-qc/9404013 at *xxx.lanl.gov*, Apr. 8, 1994.

22. In a later revision of their paper, Gell-Mann and Hartle do not speak so specifically about "communicating" and instead refer to "drawing inferences" about IGUSes in a different quasiclassical realm.

23. Murray Gell-Mann and James Hartle, "Equivalent Sets of Histories and Multiple Quasiclassical Domains," as revised May 5, 1996.

24. Roland Omnès, "Consistent Interpretations of Quantum Mechanics," *Reviews of Modern Physics* 64 (Apr. 1992) 339–82 and *The Interpretation of Quantum Mechanics* (Princeton, N.J.: Princeton University Press, 1994).

25. Roland Omnès, "Consistent Interpretations of Quantum Mechanics," 359. Some physicists do not believe that any of the quantum decoherence approaches have solved all the problems posed by quantum physics. Adrian Kent of the University of Cambridge has written several papers criticizing aspects of Gell-Mann and Hartle's approach. Another critic is Anthony Leggett, a quantum physicist at the University of Illinois at Urbana-Champaign. "This is all dressed up in all sorts of fancy words, like decoherence, but to my mind it really doesn't touch the basic issue," he told me in an interview. Analyzing the interaction with the environment may explain which definite result emerges, but doesn't really explain why quantum mechanics permits a definite result to happen. "The crucial question is how can the quantum . . . [mathematics] ever give you any definite outcome at all? Not just the particular outcome that I saw, but how did I get a definite outcome at all? . . . You have to go outside . . . quantum mechanics and postulate that quantum mechanics is not the whole truth," Leggett said.

26. Omnès, "Consistent Interpretations of Quantum Mechanics," 364.

27. Wojciech Zurek, interview in Los Alamos, N.M., Apr. 26, 1993.

28. Ibid.

29. J. J. Halliwell, "Somewhere in the Universe: Where Is Information Stored When Histories Decohere?" preprint quant-ph/9902008 at *xxx.lanl.gov*, Feb. 2, 1999, 10.

30. All quotes in the discussion of Hugh Everett's "many-worlds" theory are taken from his articles in Bryce DeWitt and Neill Graham, eds., *The Many-Worlds Interpretation of Quantum Mechanics* (Princeton, N.J.: Princeton University Press, 1973).

Chapter 10: From Black Holes to Supermatter

1. John Wheeler, *A Journey into Gravity and Spacetime* (New York: Scientific American Library, 1990), 211.
2. John Wheeler, interview in Princeton, Feb. 10, 1998.
3. The Sigma Xi-Phi Beta Kappa annual lecture at the American Association for the Advancement of Science meeting, Dec. 29, 1967, in New York City.
4. John Wheeler, "Our Universe: The Known and the Unknown," *American Scientist* 56 (Spring 1968): 9.
5. For this phrase I owe my old editor Jim Browder, who used it in a headline he wrote on a story I did in 1978.
6. Leonard Susskind, colloquium in Dallas, Dec. 7, 1992.
7. Ibid.
8. John Wheeler, *A Journey into Gravity and Spacetime*, 221.
9. John Wheeler, interview at Dallas–Fort Worth International Airport, Apr. 14, 1990.
10. John Wheeler, *A Journey into Gravity and Spacetime*, 221.
11. Hawking is often compared to Einstein in many ways, including achievements in physics. I think this goes too far. Hawking has a brilliant mind, but he has done nothing remotely as significant as what Einstein did in terms of revolutionizing the entire human conception of reality and the universe. I don't mean this in any way as a slam on Hawking—it should be no insult to say somebody is not as great as Einstein was. But Einstein stood above all other physicists of his era, and perhaps any era. Many other physicists working today have accomplished as much as Hawking has.
12. It's called the Texas symposium because the first one, in 1963, was held in Dallas.
13. News conference at the Texas Symposium on Relativistic Astrophysics, Berkeley, Calif., Dec. 15, 1992.
14. A couple of years later Wilczek was advocating a different approach. When I was visiting the Institute for Advanced Study in 1995, he explained to me that Hawking's calculations, showing that radiation from a black hole would be random, assumed a fixed geometry of space at the black hole's boundary. Actually, quantum mechanics requires the geometry of space at small scales to be in constant flux. "Instead of space and time being a fixed thing, instead of having one definite space and time, you have coherent superpositions of many different spaces and times," Wilczek said. Taking this flux of space into consideration, he and Per Kraus of Princeton University published a paper in 1994 showing that the outgoing Hawking radiation might not be entirely random after all, and might therefore contain the information thought to be lost.
15. Tom Siegfried, "Strings may stop black holes from shedding information," *Dallas Morning News*, Oct. 25, 1993, 9D.
16. Christopher Fuchs, Workshop on Physics and Computation, Dallas, Oct. 2, 1992.

17. Christopher Fuchs, "Landauer's Principle and Black-Hole Entropy," Workshop on Physics and Computation: PhysComp '92 (Los Alamitos, Calif.: IEEE Computer Society Press, 1993), 91.

18. Jeremy Gray, *Ideas of Space* (Oxford: Clarendon Press, 1989), 123.

19. This gets really complicated, of course, because there is always a question of who is moving rapidly with respect to whom, and returning from a trip requires a spacecraft to make some turns, or accelerations, that complicate the analysis. A good account of all these details can be found in Paul Davies's book *About Time* (New York: Simon & Schuster, 1995).

20. Hermann Weyl, *Symmetry* (Princeton, N.J.: Princeton University Press, 1952), 5.

21. John Schwarz, interview in Pasadena, Calif., Oct. 22, 1996.

22. I wrote about the WIMP sighting as a possible discovery in my Monday column. James Glanz of the journal *Science* was at the meeting and also wrote about the possible WIMP discovery.

23. Edward Witten, interview in Princeton, N. J., Feb. 11, 1998.

Chapter 11: The Magical Mystery Theory

1. Tom Siegfried, "The space case," *Dallas Morning News,* Oct. 9, 1995, 6D.

2. Nathan Seiberg, interview in Princeton, N.J., Oct. 21, 1998.

3. An important implication of general relativity is that the space of the universe is not constant but is always changing in extent. Einstein's equations say the universe has to be either getting bigger or getting smaller. Right now everybody agrees that it is getting bigger—that the universe is expanding.

4. Pawel Mazur of the University of South Carolina and colleagues have written papers making this point. "We have advanced the proposition that . . . at cosmological distance scales quantum fluctuations of the metric can be important and modify drastically the classical metric description of general relativity." Ignatios Antoniadis, Pawel Mazur and Emil Mottola, "Physical States of the Quantum Conformal Factor," hep-th/9509169 at *xxx.lanl.gov.*

5. Lee Smolin, interview in Princeton, N.J., May 9, 1995.

6. John Schwarz, interview in Pasadena, Calif., Oct. 22, 1996.

7. Edward Witten, "The holes are defined by the string," *Nature* 383 (Sept. 19, 1996), 215–16.

8. A very painful trip, because of an ingrown toenail. I had to limp around the campus with a shoe half off my foot. The whole story of my toe problems cannot be told here; it would take another whole book.

9. Michael Duff, interview in College Station, Tex., Oct. 14, 1996.

10. Joseph Polchinski, telephone interview, Oct. 16, 1996.

11. Andrew Strominger, telephone interview, Oct. 22, 1996.

12. John Schwarz, interview, Oct. 22, 1996.

13. Ibid.

14. Nathan Seiberg, interview in Princeton, N.J., Oct. 21, 1998.

15. The D-brane approach, which works on particular types of black holes, gets the right value of one-fourth the surface area; the loop approach gets the right answer in that the entropy is proportional to the surface area, but does not precisely predict the factor of one-fourth.

16. Abhay Ashtekar et al., "Quantum Geometry and Black Hole Entropy," *Physical Review Letters* 80 (Feb. 2, 1998), 907.

Chapter 12: The Bit and the Pendulum

1. Perhaps a better phrase would be "representation of reality," for I don't wish to enter into the argument over whether science should be only descriptive and not explanatory—that is another issue.

2. Hugh Everett, "Remarks on the Role of Theoretical Physics," in Bryce De-Witt and Neill Graham, eds., *The Many-Worlds Interpretation of Quantum Mechanics* (Princeton, N.J.: Princeton University Press, 1973), 133.

3. Ibid., 134.

4. Ibid.

5. David Landes, *Revolution in Time* (Cambridge, Mass.: Harvard University Press, 1983), 118.

6. Harmonic oscillators have to do with much more than just music, although many musical instruments make good examples. Most common objects become harmonic oscillators when they are disturbed from a stable state at rest (equilibrium) and oscillate back and forth around that stability point, just the way a pendulum swings back and forth. So the analysis of pendulum motion can be used to study the motion of many objects in nature. Musical instruments work by making such "harmonic" oscillations, whether a vibrating string as in a violin or vibrating membrane (drum) or vibrating air in a clarinet.

7. Stephen Toulmin, "From Clocks to Chaos: Humanizing the Mechanistic World View," in Hermann Haken, Anders Karlqvist, and Uno Svedin, eds., *The Machine as Metaphor and Tool* (Berlin: Springer-Verlag, 1993), 152.

8. Norbert Wiener, *The Human Uses of Human Beings*, (1954; reprint New York: Da Capo Press, 1988), 15.

9. Peter Galison, "The Americanization of Unity," *Daedalus* 127 (Winter 1998); 59.

10. Rolf Landauer, lecture in Columbus, Ohio, Apr. 21, 1998.

11. I think there's a similarity between Landauer's view and some discussions of the arrow of time as related to quantum decoherence. In the decoherence process, the transfer of information from a system to its environment seems to be irreversible. "This has to do with being able to lose information, having a way information can flow away rather than be coming in at the same time," Andy Albrecht explained to me. Irreversible loss of energy, in accord with the second law of thermodynamics, accompanies the loss of information to the environment, so the quantum arrow of time corresponds to the thermodynamic arrow of time. And as Roland Omnès has pointed

out, even though quantum mathematics seems to permit conditions that would make it possible to reverse the direction of time, it would be an immense task to prepare the proper conditions. It might, for example, require an apparatus so large that the universe could not contain it. This strikes me as similar to Landauer's concern that computing resources would be limited by the physical nature of the universe itself.

12. Again, this seems to me to fit nicely with research on this issue from other perspectives. In fact, the Gell-Mann–Hartle approach of decoherent histories requires some intrinsic coarse graining in the quantum world. My personal speculation is that something like Landauer's view might someday help provide a sturdy foundation for the decoherence-consistent histories approach to quantum mechanics.

13. John Wheeler, "Law without Law," in *Frontiers of Time* (Austin, Tex.: Center for Theoretical Physics, University of Texas at Austin, 1978), 3.

14. John Wheeler, "Information, Physics, Quantum: The Search for Links," in Anthony J. G. Hey, ed., *Feynman and Computation* (Reading, Mass.: Perseus Books, 1999), 314.

15. And he continued: "When December 2000 rolls around and we all celebrate the hundreth anniversary of Planck's great discovery, if I am invited to give a talk I think I will title it 'A Hundred Years of the Quantum—the Glory and the Shame'—glory because of all the wonderful mechanisms the quantum has allowed us to understand and shame because we still don't know where the principle itself comes from."

16. John Wheeler, "Information, Physics, Quantum: The Search for Links," in Anthony J. G. Hey, ed., *Feynman and Computation* (Reading, Mass.: Perseus Books, 1999). Reproduced from the Proceedings of the Third International Symposium on the Foundations of Quantum Mechanics, Tokyo, 1989, 311.

17. Ibid.

18. Ibid.

19. Ibid., 313.

20. This is because, in its unsimplified form, the surface area formula for a black hole contains Planck's constant, the critical quantity of quantum theory. Planck's constant expresses the relationship between the frequency of radiation and the amount of energy that one quantum of that radiation contains.

21. John Wheeler, "Information, Physics, Quantum: The Search for Links," in Anthony J. G. Hey, ed., *Feynman and Computation* (Reading, Mass.: Perseus Books, 1999). Reproduced from the Proceedings of the Third International Symposium on the Foundations of Quantum Mechanics, Tokyo, 1989, 313.

22. Robert Wald, "Gravitation, Thermodynamics and Quantum Theory," preprint gr-qc/9901033 at *xxx.lanl.gov,* Jan. 12, 1999, 2.

Glossary

algorithm   A set of rules describing a step-by-step process for the solution of a problem.

algorithmic information content (or algorithmic complexity, Kolmogorov complexity)   A measure of the shortest computer program that can produce a message. The more random a message, the longer the computer program that reproduces it must be. Technically, the algorithmic complexity is the number of bits in the smallest program that outputs the string of bits composing the message when that program is run on a universal Turing machine.

anthropic principle   The idea that life in some way or another bears a special relationship to the universe. In its weak form, the anthropic principle states that the physical features of the universe must be such that life can exist. In its strong form, the anthropic principle states that the universe itself cannot exist without the presence of life.

black hole   A region of space with a gravitational field so intense that nothing can escape from within it. Black holes may be produced by the collapse of a massive star at the end of its lifetime or by the agglomeration of large masses at the center of a galaxy.

ciliates   Microscopic organisms covered with miniature whiplike appendages called cilia that aid in locomotion. Ciliates typically contain two cell nuclei.

consistent histories   An interpretation of **quantum mechanics** holding that the range of future possible events in a quantum system is constrained by past recorded events.

Copenhagen interpretation   An interpretation of **quantum mechanics,** developed in the 1920s by Niels Bohr. The central feature of the Copenhagen interpretation is Bohr's principle of complementarity, which holds that quantum systems can be described in two mutually exclusive, yet complementary, ways when the entire experimental arrangement is taken into account. For example, an electron can in some situations behave as a particle, but in others behave as a wave, depending on the nature of the experiment being performed.

decidability question   A question posed by mathematicians early in the twentieth century about whether a guaranteed method could be devised for deciding the answers to any mathematical problem. The question is equivalent to asking whether there is any automatic way of determining the truth of any proposition.

entropy   A measure of the disorder in a physical system, deeply related to the idea of information about how the parts of a system are put together.

general theory of relativity   Einstein's theory of gravity. General relativity ascribes gravitation to the warping of spacetime by an object possessing mass (or energy). The resulting curvature of spacetime guides the paths of moving objects. In other words, mass warps spacetime, telling it how to curve, and spacetime grips mass, telling it how to move.

G protein   A complex of three protein subunits found within cells, important in many life processes. G proteins typically relay messages received outside the cell to other components inside the cell, inducing appropriate chemical responses.

Hawking radiation   Radiation emitted from near the surface of a **black hole.** Even though nothing inside a black hole can truly escape, Hawking radiation drains mass from a black hole, causing it to "evaporate" over long time scales. Named for Stephen Hawking.

Landauer's principle   The requirement that energy is used up, or dissipated, any time that a bit of information is erased. Named for Rolf Landauer.

laws of thermodynamics   The fundamental laws of physics describing the flow and exchange of heat.

first law of thermodynamics   The same thing as the law of conservation of energy—the total amount of energy in any closed system remains fixed.

second law of thermodynamics   The requirement that natural processes in a closed system seek a state of equilibrium; equivalent to the idea that disorder always increases over time in a closed system and that energy used in doing work cannot be reused.

logical depth   A measure of organization in a system, essentially the number of steps in a deductive or causal path connecting the system with its plausible origin. In computational terms, the logical depth could be expressed as the time required by a universal computer to compute the system in question from a program that could not itself have been computed from a more concise program.

many-worlds interpretation   An interpretation of **quantum mechanics** holding that all the multiple possibilities in a quantum system have an equal claim to being real.

Maxwell's demon   A hypothetical being who could direct the motion of molecules in a way so as to circumvent the **second law of thermodynamics.**

photoelectric effect   The property by which certain substances emit electrons when struck by light.

photon   A particle of light or other form of electromagnetic radiation, with an energy related to the frequency of the radiation as required by quantum theory. Technically, a quantum of electromagnetic radiation.

Planck's constant   The numerical quantity relating the frequency of radiation to the energy possessed by one quantum of that radiation. Named for Max Planck.

quantum decoherence   The process by which multiple possible realities permitted by **quantum mechanics** are eliminated when a system interacts with its environment.

quantum mechanics   The mathematical framework governing the micropscopic world and specifying the behavior of matter and energy in fundamental processes. While its strangest features are apparent mostly on the subatomic level, quantum mechanics applies, in principle, to the entire universe.

qubit   A bit of quantum information. Unlike an ordinary bit of information, which can be expressed as either a 0 or a 1, a quantum bit is typically a combination of 0 and 1. Pronounced CUE-bit.

Shannon information or **Shannon entropy**   A measure of the uncertainty resolved by a message—or, from the sender's point of view, a measure of the freedom of choice in composing a message. If there are only two possible messages (say, yes or no), then sending one reduces the receiver's uncertainty by one bit, the unit in which Shannon information is measured. Technically, the number of bits is equal to the negative logarithm of the probability of a message.

special theory of relativity   Einstein's first theory of relativity, describing the laws of motion independently of the uniform motion of any observer. Special relativity requires the speed of light to be the same regardless of the motion of who's measuring it, and one of the consequences is the equivalence of mass and energy.

spin networks   Representations of space in the form of graphs connecting lines to points. The points represent the angular momentum, or "spin," of elementary particles, and the lines represent the paths of the particles. An idea devised by the British mathematician Roger Penrose in the 1970s and now used by some researchers studying the nature of gravity.

string theory   An approach to physical theory that regards the fundamental entities not as pointlike particles, but as one-dimensional extended objects, or strings.

superposition   The quantum-mechanical concept of multiple possible states of a system existing simultaneously. In the classic example, a hypothetical cat in a box, possibly poisoned or possibly not, is in a superposition of dead-and-alive.

Turing machine   An idealized computer, or any machine, that could implement an **algorithm** to solve a mathematical problem. A **universal Turing machine** could emulate any other Turing machine, and therefore solve any problem that could be solved by the implementation of an algorithm.

uncertainty principle   The principle devised by Werner Heisenberg that forbids the simultaneous measurement of certain properties of an object or elementary particle such as an electron. The position and momentum of such a particle cannot both be measured precisely, for example, nor could the energy of a particle be specified for an arbitrarily short duration of time. Though often regarded as a consequence of the process of measurement, the uncertainty principle actually just describes the fact of **quantum mechanics** that an elementary particle does not possess a precise position or momentum at any point in time.

wormholes   Hypothetical tubes leading from one point in spacetime to another without passing through the intervening space. Wormholes of submicroscopic size may be a feature of the vacuum of space at extremely small scales.

Further Reading

The following list includes many primary sources for the ideas described in this book, as well as some secondary sources, review articles, and popular accounts, including several articles I have written for the *Dallas Morning News*. (Those articles are available for a nominal fee at the *Morning News* World Wide Web archive, *http://archive. dallasnews.com.*)

Introduction
Wheeler, John. "Information, Physics, Quantum: The Search for Links." In Anthony J. G. Hey, ed., *Feynman and Computation*. Reading, Mass.: Perseus Books, 1999. Reproduced from the Proceedings of the Third International Symposium on the Foundations of Quantum Mechanics, Tokyo, 1989, 354–68.

_____. *Geons, Black Holes, and Quantum Foam: A Life in Physics.* New York: W. W. Norton, 1998.

Chapter 1: Beam Up the Goulash
Bouwmeester, Dik, et al. "Experimental Quantum Teleportation." *Nature* 390 (Dec. 11, 1997): 575–79.

Ekert, Artur. "Shannon's Theorem Revisited." *Nature* 367 (Feb. 10, 1994): 513–14.

Siegfried, Tom. "Breaking quantum code offers way of sending secret messages." *Dallas Morning News*, Aug. 19, 1991.

_____. "Quantum particles may become a useful communications tool." *Dallas Morning News*, Mar. 29, 1993.

_____. "Beyond bits: emerging field hopes to exploit quantum quirkiness in information processing, computing." *Dallas Morning News*, June 20, 1994.

_____. "Quantum cryptographers claim they know how to keep a secret." *Dallas Morning News*, May 12, 1997.

Chapter 2: Machines and Metaphors
Babbage, Charles. *Passages from the Life of a Philosopher.* Edited by Martin Campbell-Kelly. New Brunswick, N.J.: Rutgers University Press, 1994.

Bernstein, Jeremy. *The Analytical Engine.* New York: Random House, 1964.

Blohm, Hans, Stafford Beer, and David Suzuki. *Pebbles to Computers: The Thread*. Toronto: Oxford University Press, 1986.

Carnot, Sadi. *Reflections on the Motive Power of Heat*. Trans. R. H. Thurston. New York: American Society of Mechanical Engineers, 1943.

Cardwell, D.S.L. *Turning Points in Western Technology*. New York: Science History Publications, 1972.

_____. *From Watt to Clausius*. Ames, Iowa: Iowa State University Press, 1989.

Casti, John. "Computing the Uncomputable." *New Scientist* (May 17, 1997): 30–33.

Goldstine, Herman. *The Computer: From Pascal to von Neumann*. Princeton, N.J.: Princeton University Press, 1972, 1993.

Haken, Hermann, Anders Karlqvist, and Uno Svedin, eds. *The Machine as Metaphor and Tool*. Berlin: Springer-Verlag, 1993.

Hey, Anthony J. G., ed. *Feynman and Computation*: Reading, Mass.: Perseus Books, 1999.

Hodges, Andrew. *Alan Turing: The Enigma*. New York: Simon & Schuster, 1983.

Macrae, Norman, *John von Neumann*. New York: Pantheon Books, 1991.

Whitrow, G. J. *Time in History*. New York: Oxford University Press, 1988.

Chapter 3: Information Is Physical

Atkins, P. W. *The Second Law*. New York: W. H. Freeman, 1984.

Bennett, Charles H. "Dissipation, Information, Computational Complexity and the Definition of Organization," in David Pines, ed. *Emerging Syntheses in Science*. Redwood City, Calif.: Addison-Wesley, 1988.

Landauer, Rolf. "Information Is Physical." *Physics Today* (May 1991): 23–29.

_____. "Information Is Inevitably Physical," in A.J.G. Hey, ed. *Feynman and Computation*. Reading, Mass.: Perseus, 1999.

Leff, Harvey S., and Andrew F. Rex, eds. *Maxwell's Demon*. Princeton, N.J.: Princeton University Press, 1990.

Pierce, John. *An Introduction to Information Theory*. 2d ed. New York: Dover Publications, 1980.

Shannon, Claude, and Warren Weaver. *The Mathematical Theory of Communication*. Urbana, Ill.: University of Illinois Press, 1949.

Siegfried, Tom. "Computer scientists look to reverse logic." *Dallas Morning News*, Apr. 19, 1993.

Zurek, W. H. "Algorithmic Information Content, Church-Turing Thesis, Physical Entropy, and Maxwell's Demon," in W. H. Zurek, ed., *Complexity, Entropy and the Physics of Information*. Redwood City, Calif.: Addison-Wesley, 1990.

Zurek, Wojciech. "Decoherence and the Transition from Quantum to Classical." *Physics Today* 44 (Oct. 1991): 36–44.

Chapter 4: The Quantum and the Computer

Brassard, Gilles. "New Trends in Quantum Computing." Preprint quant-ph/9602014 at *xxx.lanl.gov,* Feb. 19, 1996, 1.

DiVincenzo, David. "Quantum Computation." *Science* 270 (Oct. 13, 1995), 255–61.

Gershenfeld, Neil, and Isaac Chuang. "Quantum Computing with Molecules." *Scientific American* 278 (June 1998): 66–71.

Lloyd, Seth. "Quantum-Mechanical Computers." *Scientific American* 273 (Oct. 1995), 140–45.

Siegfried, Tom. "Computers encounter novel world of quantum mysteries." *Dallas Morning News,* Aug. 3, 1992.

_____. "Quantum leap in computing could make big future impact." *Dallas Morning News,* Sept. 27, 1993.

_____. "Beyond bits: emerging field hopes to exploit quantum quirkiness in information processing, computing." *Dallas Morning News,* June 20, 1994.

_____. "Quantum devices could compute where no one's computed before." *Dallas Morning News,* June 20, 1994.

_____. "Bits of progress: recent advances bring theoretical quantum computer closer to becoming a reality." *Dallas Morning News,* Feb. 12, 1996.

_____. "Computers poised for a quantum leap." *Dallas Morning News,* Mar. 16, 1998.

Chapter 5: The Computational Cell

Adleman, Leonard M. "Computing with DNA." *Scientific American* 279 (Aug. 1998): 54–61.

_____. "Molecular Computation of Solutions to Combinatorial Problems." *Science* 266 (Nov. 11, 1994): 1021–24.

Kauffman, Stuart. *At Home in the Universe.* New York: Oxford University Press, 1995.

Ross, Elliott. "Twists and Turns on G-protein Signalling Pathways." *Current Biology* 2 (1992): 517–19.

Siegfried, Tom. "Computing with chemicals." *Dallas Morning News,* Feb. 3, 1992.

_____. "For the record, DNA helps scientists track humans' steps." *Dallas Morning News,* Nov. 8, 1993.

_____. "Information-age scientists say cells' role as computer adds up." *Dallas Morning News,* Aug. 21, 1995.

_____. "DNA could be ultimate answer in long chain of computing ideas." *Dallas Morning News,* June 16, 1997.

Chapter 6: The Computational Brain

Black, Ira. *Information in the Brain.* Cambridge, Mass.: MIT Press, 1991.

Churchland, Patricia S., and Terrence J. Sejnowski. *The Computational Brain.* Cambridge, Mass.: MIT Press, 1992.

Koch, Christof. *Biophysics of Computation.* New York: Oxford University Press, 1999.

Siegfried, Tom. "Plugging in to how the brain computes." *Dallas Morning News,* Mar. 22, 1993.

Von Neumann, John. *The Computer and the Brain.* New Haven, Conn.: Yale University Press, 1958.

Chapter 7: Consciousness and Complexity

Edelman, Gerald. *Bright Air, Brilliant Fire.* New York: Basic Books, 1992.

Gazzaniga, Michael. *Nature's Mind.* New York: Basic Books, 1992.

Siegfried, Tom. "'Blindsight' brain scans make science aware of consciousness." *Dallas Morning News,* Sept. 1, 1997.

_____. "Abilities without awareness are commentary on consciousness." *Dallas Morning News,* Sept. 8, 1997.

_____. "Computing approaches of nature, machines may someday converge." *Dallas Morning News,* Sept. 22, 1997.

_____. "Outlook provides an alternative to computer model of the brain." *Dallas Morning News,* May 25, 1998.

_____. "Dynamic thought process could also have structured computing." *Dallas Morning News,* June 1, 1998.

_____. "If computers create chaos, maybe chaos can compute." *Dallas Morning News,* Sept. 28, 1998.

Tononi, Giulio, and Gerald Edelman. "Consciousness and Complexity." *Science* 282 (Dec. 4, 1998): 1846–51.

Weiskrantz, Lawrence. *Consciousness Lost and Found.* New York: Oxford University Press, 1997.

Chapter 8: IGUSes

Crutchfield, James. "The Calculi of Emergence: Computation, Dynamics and Induction." *Physica D* 75 (1994): 11–54.

Deutsch, David. "Quantum Computation." *Physics World* (June 1992): 57–61.

Gell-Mann, Murray. *The Quark and the Jaguar.* New York: W. H. Freeman, 1994.

Siegfried, Tom. "Sorting quantum from classical can be fuzzy business in physics." *Dallas Morning News,* May 30, 1994.

_____. "Overlapping domains are among universe's realms of possibilities." *Dallas Morning News,* June 6, 1994.

Waldrop, M. Mitchell. *Complexity.* New York: Simon & Schuster, 1992.

Zurek, W.H., ed. *Complexity, Entropy and the Physics of Information.* Redwood City, Calif.: Addison-Wesley, 1990

Chapter 9: Quantum Reality

Barrow, John D., and Frank J. Tipler. *The Anthropic Cosmological Principle.* New York: Oxford University Press, 1986.

Davies, P.C.W., and J. R. Brown, eds. *The Ghost in the Atom.* Cambridge, Mass.: Cambridge University Press, 1986.

DeWitt, Bryce. "Quantum Mechanics and Reality." *Physics Today* 23 (Sept. 1970): 30–35.

Everett, Hugh. "The Theory of the Universal Wave Function," in Bryce DeWitt and Neill Graham, eds., *The Many-Worlds Interpretation of Quantum Mechanics.* Princeton, N.J.: Princeton University Press, 1973.

Everett, Hugh. "'Relative State' Formulation of Quantum Mechanics." *Reviews of Modern Physics* 29 (July 1957), in Bryce DeWitt and Neill Graham, eds., *The Many-Worlds Interpretation of Quantum Mechanics.* Princeton, N.J.: Princeton University Press, 1973.

Gell-Mann, Murray. *The Quark and the Jaguar.* New York: W. H. Freeman, 1994.

Gell-Mann, Murray, and Hartle, James. "Quantum Mechanics in the Light of Quantum Cosmology," in Zurek, W. H., ed., *Complexity, Entropy and the Physics of Information.* Redwood City, Calif.: Addison-Wesley, 1990.

Griffiths, Robert B. "Consistent Interpretation of Quantum Mechanics Using Quantum Trajectories." *Physical Review Letters* 70 (Apr. 12, 1993): 2201–4.

Omnès, Roland. "Consistent Interpretations of Quantum Mechanics." *Reviews of Modern Physics* 64 (1992): 339–82.

――――. *The Interpretation of Quantum Mechanics.* Princeton, N.J.: Princeton University Press, 1994.

Siegfried, Tom. "Investigating subatomic oddities: experiments refute Einstein." *Dallas Morning News,* June 16, 1986.

――――. "Feline physics: cat paradox revives theory of many parallel universes." *Dallas Morning News,* Oct. 19, 1987.

――――. "Quantum truth is stranger than fiction." *Dallas Morning News,* Oct. 19, 1987.

――――. "Wondering whether the universe exists." *Dallas Morning News,* Mar. 14, 1988.

――――. "Spectators influence the results in quantum version of baseball." *Dallas Morning News,* Mar. 16, 1990.

――――. "Bell's quantum insight leaves a mystery for next generation." *Dallas Morning News,* Oct. 15, 1990.

――――. "The small picture: 'decoherence' may be putting together pieces of quantum puzzle." *Dallas Morning News,* Nov. 8, 1993.

Wheeler, John A. "The Universe as Home for Man," *American Scientist* 62 (1974): 683–91.

Zeh, H. D. *The Physical Basis for the Direction of Time.* Heidelberg: Springer, 1989.

Zurek, Wojciech. "Decoherence and the Transition from Quantum to Classical." *Physics Today* 44 (Oct. 1991): 36–44.

Chapter 10: From Black Holes to Supermatter

Ashtekar, Abhay, et al. "Quantum Geometry and Black Hole Entropy." *Physical Review Letters* 80 (Feb. 2, 1998): 904–7.

Landauer, Rolf. "Computation, Measurement, Communication and Energy Dissipation." In S. Haykin, ed., *Selected Topics in Signal Processing*. Englewood Cliffs, N.J.: Prentice-Hall, 1989.

Siegfried, Tom. "Black holes pose challenge." *Dallas Morning News*, Dec. 28, 1992.

_____. "Supersymmetry could impose order beyond Einstein's vision." *Dallas Morning News*, Jan. 1, 1996.

Wheeler, John. "Information, Physics, Quantum: The Search for Links." In Anthony J. G. Hey, ed., *Feynman and Computation*. Reading, Mass.: Perseus Books, 1999. Reproduced from the Proceedings of the Third International Symposium on the Foundations of Quantum Mechanics, Tokyo, 1989, 354–68.

Wheeler, John Archibald. *Geons, Black Holes, and Quantum Foam*. New York: W. W. Norton, 1998.

_____. "Time Today." In J. J. Halliwell et al., eds., *Physical Origins of Time Asymmetry*. Cambridge, Mass.: Cambridge University Press, 1994.

Wheeler, John. *A Journey into Gravity and Spacetime*. New York: Scientific American Library, 1990.

Chapter 11: The Magical Mystery Theory

Newman, James R. Introduction to *The Common Sense of the Exact Sciences*, by William Kingdom Clifford. New York: Dover, 1955.

Siegfried, Tom. "Loop view of space's geometry shows promise in physics circles." *Dallas Morning News*, May 15, 1995.

_____. "Magic with black holes, strings revises physicists' view of space." *Dallas Morning News*, June 26, 1995.

_____. "The space case." *Dallas Morning News*, Oct. 9, 1995.

_____. "Hawking votes for bubbles, gives wormholes the hook." *Dallas Morning News*, Nov. 6, 1995.

_____. "Physicists sing praises of magical mystery theory." *Dallas Morning News*, Oct. 28, 1996.

_____. "Mother of all theories needs help from two-timing father." *Dallas Morning News*, Oct. 28, 1996.

Witten, Edward. "The holes are defined by the string." *Nature* 383 (Sept. 19, 1996): 215–16.

Chapter 12: The Bit and the Pendulum

Galison, Peter. "The Americanization of Unity." *Daedalus* 127 (Winter 1998): 45–71.

Toulmin, Stephen. "From Clocks to Chaos: Humanizing the Mechanistic World View." In Hermann Haken, Anders Karlqvist, and Uno Svedin, eds., *The Machine as Metaphor and Tool*. Berlin: Springer-Verlag, 1993.

Weyl, Hermann. *Philosophy of Mathematics and Natural Science*. Princeton, N.J.: Princeton University Press, 1949.

Index